HIERARCHICAL MODELING FOR VLSI CIRCUIT TESTING

THE KLUWER INTERNATIONAL SERIES
IN ENGINEERING AND COMPUTER SCIENCE
VLSI, COMPUTER ARCHITECTURE AND
DIGITAL SIGNAL PROCESSING

Consulting Editor
Jonathan Allen

Other books in the series:

Logic Minimization Algorithms for VLSI Synthesis. R.K. Brayton, G.D. Hachtel, C.T. McMullen, and Alberto Sanngiovanni-Vincentelli. ISBN 0-89838-164-9.
Adaptive Filters: Structures, Algorithms, and Applications. M.L. Honig and D.G. Messerschmitt. ISBN 0-89838-163-0.
Introduction to VLSI Silicon Devices: Physics, Technology and Characterization. B. El-Kareh and R.J. Bombard. ISBN 0-89838-210-6.
Latchup in CMOS Technology: The Problem and Its Cure. R.R. Troutman. ISBN 0-89838-215-7.
Digital CMOS Circuit Design. M. Annaratone. ISBN 0-89838-224-6.
The Bounding Approach to VLSI Circuit Simulation. C.A. Zukowski. ISBN 0-89838-176-2.
Multi-Level Simulation for VLSI Design. D.D. Hill and D.R. Coelho. ISBN 0-89838-184-3.
Relaxation Techniques for the Simulation of VLSI Circuits. J. White and A. Sangiovanni-Vincentelli. ISBN 0-89838-186-X.
VLSI CAD Tools and Applications. W. Fichtner and M. Morf, Editors. ISBN 0-89838-193-2.
A VLSI Architecture for Concurrent Data Structures. W.J. Dally. ISBN 0-89838-235-1.
Yield Simulation for Integrated Circuits. D.M.H. Walker. ISBN 0-89838-244-0.
VLSI Specification, Verification and Synthesis. G. Birtwistle and P.A. Subrahmanyam. ISBN 0-89838-246-7.
Fundamentals of Computer-Aided Circuit Simulation. W.J. McCalla. ISBN 0-89838-248-3.
Serial Data Computation. S.G. Smith, P.B. Denyer. ISBN 0-89838-253-X.
Phonologic Parsing in Speech Recognition. K.W. Church. ISBN 0-89838-250-5.
Simulated Annealing for VLSI Design. D.F. Wong, H.W. Leong, C.L. Liu. ISBN 0-89838-256-4.
Polycrystalline Silicon for Integrated Circuit Applications. T. Kamins. ISBN 0-89838-259-9.
FET Modeling for Circuit Simulation. D. Divekar. ISBN 0-89838-264-5.
VLSI Placement and Global Routing Using Simulated Annealing. C. Sechen. ISBN 0-89838-281-5.
Adaptive Filters and Equalizers. B. Mulgrew, C.F.N. Cowan. ISBN 0-89838-285-8.
Computer-Aided Design and VLSI Device Development, Second Edition. K.M. Cham, S-Y. Oh, J.L. Moll, K. Lee, P. Vande Voorde, D. Chin. ISBN: 0-89838-277-7.
Automatic Speech Recognition. K-F. Lee. ISBN 0-89838-296-3.
Speech Time-Frequency Representations. M.D. Riley. ISBN 0-89838-298-X.
A Systolic Array Optimizing Compiler. M.S. Lam. ISBN: 0-89838-300-5.
Algorithms and Techniques for VLSI Layout Synthesis. D. Hill, D. Shugard, J. Fishburn, K. Keutzer. ISBN: 0-89838-301-3.
Switch-Level Timing Simulation of MOS VLSI Circuits. V.B. Rao, D.V. Overhauser, T.N. Trick, I.N. Hajj. ISBN 0-89838-302-1.
VLSI for Artificial Intelligence. J.G. Delgado-Frias, W.R. Moore (Editors). ISBN 0-7923-9000-8.
Wafer Level Integrated Systems: Implementation Issues. S.K. Tewksbury. ISBN 0-7923-9006-7.
The Annealing Algorithm. R.H.J.M. Otten & L.P.P.P. van Ginneken. ISBN 0-7923-9022-9.
VHDL: Hardware Description and Design. R. Lipsett, C. Schaefer and C. Ussery. ISBN 0-7923-9030-X.
The VHDL Handbook. D. Coelho. ISBN 0-7923-9031-8.
Unified Methods for VLSI Simulation and Test Generation. K.T. Cheng and V.D. Agrawal. ISBN 0-7923-9025-3.
ASIC System Design with VHDL: A Paradigm. S.S. Leung and M.A. Shanblatt. ISBN 0-7923-9032-6.
BiCMOS Technology and Applications. A.R. Alvarez (Editor). ISBN 0-7923-9033-4.
Analog VLSI Implementation of Neural Systems. C. Mead and M. Ismail (Editors). ISBN 0-7923-9040-7.
The MIPS-X RISC Microprocessor. P. Chow. ISBN 0-7923-9045-8.
Nonlinear Digital Filters: Principles and Applications. I. Pitas and A.N. Venetsanopoulos. ISBN 0-7923-9049-0.
Algorithmic and Register-Transfer Level Synthesis: The System Architect's Workbench. D.E. Thomas, E.D. Lagnese, R.A. Walker, J.A. Nestor, J.V. Rajan, R.L. Blackburn. ISBN 0-7923-9053-9.
VLSI Design for Manufacturing: Yield Enhancement. S.W. Director, W. Maly, A.J. Strojwas. ISBN 0-7923-9053-7.
Testing and Reliable Design of CMOS Circuits. N.K. Jha, S. Kundu. ISBN 0-7923-9056-3.

HIERARCHICAL MODELING FOR VLSI CIRCUIT TESTING

by

Debashis Bhattacharya
Yale University

and

John P. Hayes
The University of Michigan

KLUWER ACADEMIC PUBLISHERS
Boston/Dordrecht/London

Distributors for North America:
Kluwer Academic Publishers
101 Philip Drive
Assinippi Park
Norwell, Massachusetts 02061 USA

Distributors for all other countries:
Kluwer Academic Publishers Group
Distribution Centre
Post Office Box 322
3300 AH Dordrecht, THE NETHERLANDS

Library of Congress Cataloging-in-Publication Data

Bhattacharya, Debashis, 1961–
    Hierarchical modeling for VLSI circuit testing / by Debashis
Bhattacharya, John P. Hayes.
       p. cm. — (The Kluwer international series in engineering and
computer science. VLSI, computer architecture, and digital signal
processing.)
    ISBN 0-7923-9058-X
    1. Integrated circuits—Very large scale integration—Testing.
2. Integrated circuits—Very large scale integration—Computer
simulation.   I. Hayes, John P. (John Patrick), 1944–
II. Title.   III. Series.
TK7874.B484   1990
621.39 ′5 ′0287—dc20   89-24726
                                                                        CIP

Printed in the United States of America

To our wives

TABLE OF CONTENTS

viii

Preface

Test generation is one of the most difficult tasks facing the designer of complex VLSI-based digital systems. Much of this difficulty is attributable to the almost universal use in testing of low, gate-level circuit and fault models that predate integrated circuit technology. It is long been recognized that the testing problem can be alleviated by the use of higher-level methods in which multigate modules or cells are the primitive components in test generation; however, the development of such methods has proceeded very slowly. To be acceptable, high-level approaches should be applicable to most types of digital circuits, and should provide fault coverage comparable to that of traditional, low-level methods. The fault coverage problem has, perhaps, been the most intractable, due to continued reliance in the testing industry on the single stuck-line (SSL) fault model, which is tightly bound to the gate level of abstraction.

This monograph presents a novel approach to solving the foregoing problem. It is based on the systematic use of multibit vectors rather than single bits to represent logic signals, including fault signals. A circuit is viewed as a collection of high-level components such as adders, multiplexers, and registers, interconnected by n-bit buses. To match this high-level circuit model, we introduce a high-level *bus fault* that, in effect, replaces a large number of SSL faults and allows them to be tested in parallel. However, by reducing the bus size from n to one, we can obtain the traditional gate-level circuit and models. This provides a direct way for comparing the high-level method to conventional low-level approaches. It also makes the approach truly hierarchical–a unique and useful feature. We then develop a systematic test generation algorithm for bus faults in high-level circuits, which we have implemented in a computer program called VPODEM.

In addition to presenting a theory of hierarchical modeling and testing, the book also describes the results of experiments that apply VPODEM to various types of logic circuits. These results indicate that the high-level approach alone can produce test sets with good coverage of SSL faults. The tests sets are also smaller and require less test generation effort than traditional methods, especially when the circuits being tested are moderately or highly regular. Moreover, by using VPODEM hierarchically to generate tests for any SSL faults not covered by the high-level analysis alone, 100 percent coverage of all detectable faults can can be guaranteed for any circuit. We also report some new design-for-testability methods to make circuits more amenable to the high-level testing methodology.

This work should be of interest to circuit designers, test engineers, and others concerned with the testing of complex digital systems. To make the book self-contained, we have included a short tutorial on traditional testing problems and approaches in Chapter 1. Our high-level circuit and modeling techniques are developed in Chapter 2, along with a suitable notation for describing the behavior of hierarchical systems. The test generation algorithm and its implementation (VPODEM) are covered in Chapter 3, and an experimental evaluation of VPODEM's performance is presented. Chapter 4 proposes methods to enhance the testability of digital circuits at the high level. Chapter 5 summarizes the book and suggests some directions for future research.

Most of the material in this book was developed over the past few years as part of the first author's Ph.D. dissertation research at the University of Michigan. This research was supported by a grant from the National Science Foundation and by a fellowship from International Business Machines Corporation. We wish to express our gratitude to both of these organizations. Thanks are also due to Daniel E. Atkins, Ronald J. Lomax, John F. Meyer and Trevor N. Mudge of the University of Michigan for their comments, as well as to Virginia Folsom for secretarial assistance.

HIERARCHICAL MODELING FOR VLSI CIRCUIT TESTING

Chapter 1

INTRODUCTION

Testing for hardware faults is of major importance in ensuring reliable operation of digital circuits [Bre76,Fuj86]. A circuit must normally be tested during several phases of its production, and also while it is being used in the field, to verify that it is working according to specifications. In the last two decades, circuit design and device fabrication processes have advanced rapidly, resulting in very large-scale integrated (VLSI) circuits containing thousands or millions of transistors. The large number of components in VLSI circuits has greatly increased the importance and difficulty of testing such circuits.

VLSI has been made possible to a great extent by the development of systematic and computer-aided design (CAD) techniques that allow circuits to be designed and analyzed efficiently at various levels of abstraction. However, the techniques used to test these circuits have not developed at the same rate, leading to a gap between the capabilities of the CAD tools available for circuit design, and those available for testing. While VLSI circuits are often designed and studied at the high (register) level of abstraction, current test generation techniques usually require a more complex low-level model of the circuit to be developed at the logic or gate level. The resulting increase in the number of components in the circuit model — a gate-level model contains perhaps ten times as many components as the equivalent register-level model — tends to increase significantly the time required to generate tests for the circuit. Moreover, important simplifying features of the circuit under test, such as the presence of repetitive subcircuits, must usually be ignored for test generation purposes.

This book addresses the problem of testing large circuits at a high level of abstraction to reduce the complexity of the testing problem. A new high-level circuit modeling methodology and a new class of high-level fault models are introduced. A hierarchical testing technique is also developed that uses the proposed circuit and fault models. These models and the associated testing techniques are shown to be more efficient than, but completely compatible with, classical gate-level approaches.

This chapter introduces the general testing problem, and surveys relevant prior work in algorithmic approaches to test generation. Section 1.1 defines the basic terminology and testing concepts used, classifies existing test methods and problems, and outlines the proposed methodology. Section 1.2 surveys and evaluates existing testing techniques, while Section 1.3 gives a brief outline of the book.

1.1 BACKGROUND

Digital system testing deals with hardware *faults*, which are improper states of a digital system resulting from failure of components, physical interference from the environment, operator errors, or incorrect design [Avi75]. A manifestation of a fault in the form of an incorrect (output) signal in response to a set of input signals is termed an *error*. The two major goals of testing are fault detection, and fault location or isolation. A set of input signals designated to detect or locate some faults of interest is termed a *test pattern*; a *test* is a sequence of one or more test patterns. A test is said to *detect* a fault if the responses of the circuit to the test with and without the fault present differ from each other. A test *locates* a fault if the responses also indicate the location or identity of the fault. Faults can be classified into two broad categories: permanent and transient. *Permanent* faults represent physical defects that are stable and continuous. In many cases, permanent faults are due to irreversible physical changes or design errors in the circuit under consideration. *Transient* faults, on the other hand, are temporary changes in the system's states, which are only present occasionally, and are due to unstable hardware or temporary environmental conditions such as power supply fluctuations. Faults of this type that occur repeatedly are termed *intermittent*. In this work, we will only address the detection of permanent faults.

Figure 1.1 shows the three basic steps in testing a digital circuit: test pattern generation, test pattern application, and response verification. The testing process is often implemented by computer-controlled instruments termed *automatic test equipment* (ATE) [Feu88]. Tests detecting faults in the circuit C under test are produced by a test generator and applied to C. The responses from C are verified either by comparing them with the corresponding responses of known "good" circuit, or by comparing them with separately computed "correct" responses; disagreements indicate the presence of faults. The computation of the test patterns needed to detect specified types of faults in a given circuit is a central problem in testing research. Typically, the set of test patterns is required to be as small as possible, and to detect as many of the faults of interest as possible.

The methods employed for test pattern generation are usually classified into three types: algorithmic, random and exhaustive [Bre76,Fuj86]. *Algorithmic* test generation, which is the main subject of this book, employs ana-

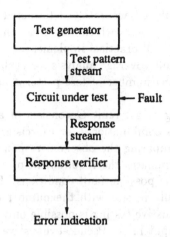

Figure 1.1: Testing a digital circuit

lytic procedures to derive test pattern sets of minimum or near-minimum size [Fuj83,Goe81,Rot66,Tho71]. It requires a well-defined logical, i.e., technology-independent, model of the circuit C under consideration, and a fault model that describes, in some convenient form, the effect of the physical faults for which tests are to be generated. The most commonly used circuit model, termed the *gate-level* model, is one that describes the circuit as an interconnection of Boolean logic gates (AND, OR, NAND, NOR, XOR, etc.). The most common fault model is the *single stuck-line* (SSL) model, which assumes that any signal line in C may be stuck at logical 1 or 0, while the circuit components (gates) are fault-free [Eld59]. The SSL model also assumes that at most one line in C is stuck-at-0 or stuck-at-1 at a time. Thus, the SSL fault model is a logical model that is applicable to the gate-level representations of all types of integrated circuits.

A given test T may detect many other faults in C besides those for which it was generated; all detected faults are said to be *covered* by T. The faults covered by a test can sometimes be determined by a fault coverage analysis, such as fault simulation, which is simpler than the test generation process itself. Often, fault coverage analysis is done concurrently with test generation to reduce the number of test patterns produced, and to avoid generating too many tests for the same fault. This can lead to a considerable reduction in the total test generation time.

Random test generation produces test sets consisting of pseudo-randomly chosen input patterns [Sav78,Sav83,Agr75]. This method can significantly reduce the computation cost of generating tests. However, it is necessary to know the number of tests that should be applied to C to provide an adequate degree of confidence that the circuit is fault-free. This number, which may be very large, can only be accurately determined from extensive analysis of the circuit behavior, a process which may be as difficult as the test generation process itself [Agr78]. All practical implementations of this approach also require some sort of fault coverage analysis for each test applied to the circuit under test, to keep the number of test patterns and the testing time within reasonable limits.

Exhaustive testing of a circuit C consists of applying all possible input patterns to C if it is combinational, or exercising fully C's state table if it is sequential, and comparing the observed responses with the expected ones [Bar83,Tan83]. It eliminates the requirement of a detailed knowledge of the circuit behavior or the possible fault conditions. However, the resultant test sets grow exponentially in size with the number of input lines to a circuit, and hence truly exhaustive testing is possible only for small circuits with few (at most 20 or so) input lines. Pseudo-exhaustive test generation techniques [Sri81a,Sri81b,McC82,Has83,Ake85] test only the major subcircuits exhaustively; again the objective is to reduce the test set size and the testing time.

Several measures, including fault coverage, testing time, and hardware overhead, are used to represent the effectiveness of different test methods. *Fault coverage* is the percentage of the faults in the circuit under test that are detected by a given set of tests [Abr83]. The popularity of the SSL fault model, especially with algorithmic test generation techniques, results from considerable experimental evidence that a test set for all SSL faults in practical circuits detects most permanent physical faults [Cas76,Fer88]. *Testing time* is usually estimated by the number of test patterns applied to the circuit C. In general, a smaller test set size implies shorter testing time, and a bigger test set size implies longer testing time. Sometimes, extra circuitry is added to C to simplify one or more phases of testing. In such cases, another relevant measure of testing cost is *circuit overhead* or *redundancy*, which is the percentage of the chip circuitry that is dedicated to facilitating testing. Typically, this is only a few percent of the total number of components (gates) in C.

Classical algorithmic test generation techniques, as mentioned earlier, use gate-level circuit and fault models. As complexity of the circuits increases, the process of test generation for all SSL faults becomes increasingly time-consuming. The structure of many circuits is greatly simplified by viewing them at the higher *register* level, where the primitive components are word gates, multiplexers, encoders, decoders, arithmetic-logic units, registers, etc., and the interconnections are multiline buses [Hay88]. Moreover, for some commercial VLSI circuits like microprocessors, the only circuit description available to the test designer is at the register level. High-level modeling and test generation, which are the focus of this book, constitute a natural step beyond the realm of classical gate-level test generation.

1.2 PRIOR WORK

This section reviews previous research into test generation algorithms and related issues in digital circuit analysis and design.

1.2.1 Test Generation for Combinational Circuits

A majority of the existing test generation algorithms for combinational circuits use a gate-level model of the circuit and the SSL fault model. The best-known of these algorithms are the D-algorithm [Rot66], PODEM [Goe81], and FAN [Fuj86]. These algorithms are based on the idea of computing an input test pattern that enables an error signal generated due to an SSL fault to propagate from the faulty line to some observable output line via some paths in the circuit.

To get a feel for the process of algorithmic test generation, we briefly review the D-algorithm and PODEM. Both these algorithms are essentially search procedures that identify a subset T of the set of possible input patterns such that applying T to the circuit under test guarantees the detection of all detectable SSL faults. The differences between the algorithms lie in the search procedures employed to identify T. While the D-algorithm explicitly keeps track of the paths propagating error signals, PODEM does so only implicitly. However, in both cases, extensive backtracking depending on the specific structure of the circuit under test, may be required to find a test pattern.

The D-Algorithm. This algorithm was defined by Roth in 1966 [Rot66]. It takes its name from the set of signal values $S = \{0, 1, X, D, \overline{D}\}$ used to describe the state of a line in the circuit during test generation. The value 0 (1) assigned to a line has the interpretation that the line carries the same signal 0 (1) in both the faulty and fault-free circuits. The assignment of D ($\overline{D}$) to a line implies that it carries the signal 1 (0) in the fault-free circuit and 0 (1) in the faulty circuit. The D and $\overline{D}$ values thus represent "discrepancy" or error signals generated by faults in the circuit. The value X denotes an unknown signal state and is typically assigned to an input line that is uninitialized.

To generate a test for a given fault F, the D-algorithm searches through the set of possible signal assignments to all the paths in a circuit such that a behavior difference between the good and faulty circuits can propagate along at least one path from the fault site to an observable output. The path thus found is called a *sensitized path*, and the input pattern needed to create it is a test pattern for F. Figure 1.2 shows an example of such a sensitized path in a gate-level logic circuit (a full adder) implementing 1-bit addition. This circuit has three input lines *a, b* and *carry-in*, and two output lines *sum* and *carry-out*; such input and output lines are commonly referred to as *primary input* and *primary output* lines. The test pattern $a, b, carry\text{-}in = 010$ for the indicated

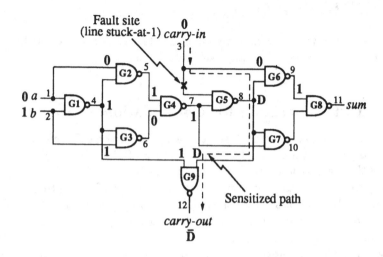

Figure 1.2: A circuit illustrating test generation.

stuck-at-1 fault creates a path (shown by a dotted line) from the fault site to the primary output line *carry-out*, such that the error signal generated by the fault ($\overline{D}$) can propagate along this path to the primary output. Thus, the indicated path is sensitized to the fault in question.

To simplify the search process, all signal assignments are made locally, i.e., to lines directly associated with paths currently under consideration, and global consistency is checked from time to time. In case of a global inconsistency, i.e., when the signal assigned to a line differs from the signal implied on that line by assignments to other lines in the circuit, the D-algorithm backtracks to some previous consistent point, and searches systematically for an alternative sensitized path. Thus the D-algorithm keeps going back and forth through the circuit until a sensitized path is found, while maintaining global consistency for all signal values assigned to the circuit.

In order to keep track of the various local signal assignments made and the various backtracking steps performed, the D-algorithm depends heavily on a systematic notation for defining the behavior of gate-level components under fault conditions [Rot66]. This notation, called the *cubical representation*, uses the set of signal values $S = \{0, 1, X, D, \overline{D}\}$ defined above. A *cube* of a component or subcircuit is an input/output value pair in which every input line is assigned a signal value from S, and every output line is assigned an appropriate signal from the same set as implied by the input assignment. Examples of cubes for a 3-input NOR gate are shown below.

$$
\begin{array}{ccccc}
1 & 1 & 0 & / & 0 \\
0 & 0 & 0 & / & 1 \\
D & 1 & D & / & 0 \\
D & D & 0 & / & \overline{D} \\
1 & X & X & / & 0 \\
X & 1 & X & / & 0 \\
X & X & X & / & X
\end{array}
\qquad (1.1)
$$

The first two cubes represent the behavior of the gate in the absence of any error signals on the inputs. For example, the cube 0 0 0 / 1 states that when all the input signals to the (fault-free) NOR gate are 0, the output signal generated by the gate must be 1. The next two (called *D-cubes*) contain D or $\overline{D}$ to represent behavior in the presence of one or more error signals on the inputs. These come in two main varieties, viz., the primitive D-cubes of a fault, and the propagation D-cubes of a gate. The *primitive D-cube of a fault* (PDCF) α in a gate G specifies the minimal input conditions for G that must be satisfied in order to produce an error signal (D or $\overline{D}$) at the output of G [Bre76,Fuj86]. A *propagation D-cube* (PDC) of gate G specifies a minimal input condition for G that must be satisfied to propagate an error signal on an input (or several inputs) to the output of G. For example, the cube 0 0 0 / D is a PDCF for a 3-input NOR gate with its output s-a-0, and D D 0/$\overline{D}$ is a PDC for a 3-input NOR gate propagating the error signals D on two of its inputs.

The fourth and fifth cubes in (1.1) show the behavior of the component when only some of the input lines have been assigned definite values (0, 1, D or $\overline{D}$). In both cases, two of the three input lines have been assigned the indefinite value X, implying that the signal assignments to those lines do not matter. In other words, these two cubes show that assigning a 1 to the first or the second input line of the gate is sufficient to produce a 0 on the output. The last cube X X X / X describes the behavior of the component when all three input lines are uninitialized. In this case, the output cannot be definitely determined, and is hence assigned the value X. Finally, note that some of the cubes in (1.1) correspond to the prime implicants of a 3-input NOR gate. Such cubes are also designated *singular cubes* and define the gate's fault-free behavior in a compact form.

Cubes are manipulated by a special type of set intersection operator by the D-algorithm. The result of *cube intersection* ($\cap$) of two cubes K_1 and K_2 of the same length is a non-empty cube $K = K_1 \cap K_2$, provided none of the corresponding pair of elements from the intersecting cubes belong to the set $\{(1,0),(0,1)\}$. The elements of K are obtained by applying the following rules for $\cap$ to corresponding elements of K_1 and K_2:

$$
s \cap s = s, \quad s \cap X = s \qquad (1.2)
$$

where $s \in \{0, 1\}$. If any corresponding pair of elements in K_1 and K_2 belong to the set $\{(1, 0), (0, 1)\}$ then $K_1 \cap K_2 = \phi$ denoting intersection of K_1 and K_2 is not possible. Some examples of cube intersection follow for the 3-input NOR case.

$$(1 \text{ X X } /0) \cap (0\ 0\ 0\ /1) = \phi$$

$$(1 \text{ X X } /0) \cap (\text{X } 1\text{X } /0) = (1\ 1 \text{ X } /0)$$

D-cube intersection is allowed even if pairs of corresponding elements of the cubes belong to the set $\{(1, 0), (0, 1)\}$, the result of intersecting 0 and 1 being defined as follows:

$$0 \cap 1 = \overline{D}, \quad 1 \cap 0 = D$$

Thus, an example of D-cube generation for the NOR gate is:

$$(1\ 1\ 0\ /0) \cap (0\ 0\ 0\ /1) = (\text{D D } 0/\overline{D})$$

Figure 1.3 summarizes the D-algorithm by means of a flowchart. Figure 1.3a shows the the overall flow of the algorithm. It also gives details of the steps followed to obtain the local assignments of values to various lines (all boxes except box 6 in Fig: 1.3a), leading to the creation of a sensitized path, if such a path exists. A special cube called the *test cube (tc)* is used to store the values of all lines in the circuit. At the beginning, all lines are uninitialized; hence the first test cube tc^0 is set to all X's in box 1. Next, the algorithm selects a PDCF for the fault in question, and thus determines a set of minimal assignments to the inputs of the faulty gate G that will result in an error signal at the output of G (box 3 in Fig. 1.3a). The intersection in box 4 (Fig. 1.3a) performs the actual assignments to the various lines in the circuit under test. This is illustrated in Fig. 1.4 when the circuit under test and the fault in question are as shown in Fig. 1.2. In this case, the faulty gate is $G5$ which is a 2-input NAND gate, and the fault is an input s-a-1 fault. Hence, the PDCF of fault 0 1 / D is intersected with tc^0 as shown Fig. 1.4 to create tc^1. Implications are next determined using the singular cubes of a 2-input NAND gate, which result in tc^2, with line 9 being assigned a 1.

The algorithm then checks for error signals on primary outputs which would indicate the completion of a sensitized path (box 5 in Fig. 1.3a). If an error signal has been propagated to one or more primary output lines, the line justification procedure (details shown in Fig. 1.3b) is invoked to perform a global consistency check. Otherwise, the algorithm tries to propagate error signals on internal lines of the circuit to primary output lines using PDC (boxes 4, 5 and 9 in Fig. 1.3a). Referring to Fig. 1.4, no error signal is found on an output line in tc^2. Hence, PDC 1 D / $\overline{D}$ is next chosen for gate $G9$ since it has a D on one of its inputs, and X on its output. The PDC is then intersected with tc^2 to generate tc^3. This leads to an error signal $\overline{D}$ on primary output *carry-out*. Line justification is performed next, as shown in Fig. 1.4 which

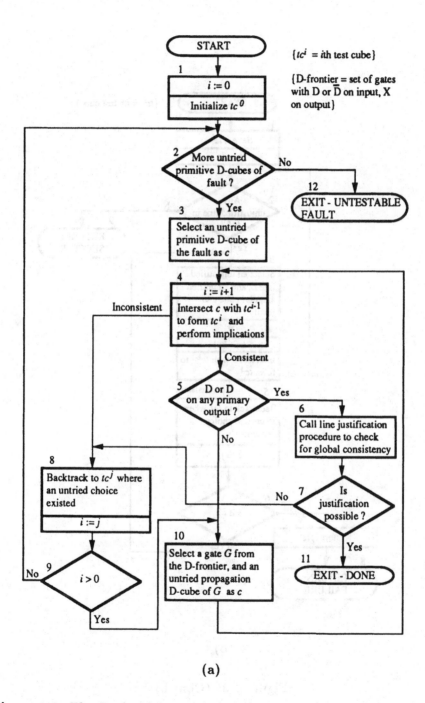

(a)

Figure 1.3: The D-algorithm: (a) main procedure; (b) line justification procedure.

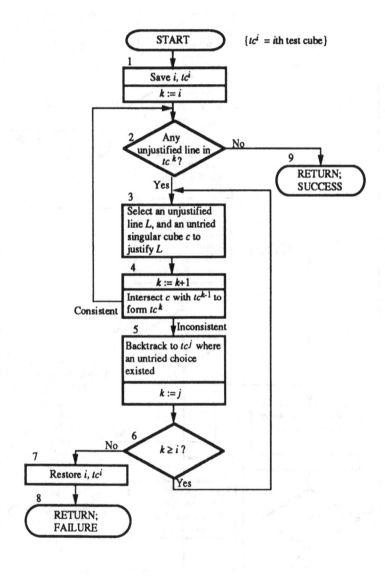

(b)

Figure 1.3: (Contd.)

	1	2	3	4	5	6	7	8	9	10	11	12	Comments
$tc^0 =$	X	X	X	X	X	X	X	X	X	X	X	X	Initialize test cube
$c =$			0				1	D					PDCF for $G5$
$tc^1 = tc^0 \cap c =$	X	X	0	X	X	X	1	D	X	X	X	X	Intersect PDCF and test cube
$c =$			0					X	1				Singular cube for $G6$
$tc^2 = tc^1 \cap c =$	X	X	0	X	X	X	1	D	1	X	X	X	Implications
$c =$				1				D				$\overline{D}$	PDC for $G9$
$tc^3 = tc^2 \cap c =$	X	X	0	1	X	X	1	D	1	X	X	$\overline{D}$	Propagate error value; sensitized path created
$c =$	0		X	1									Singular cube for $G1$
$tc^4 = tc^3 \cap c =$	0	X	0	1	X	X	1	D	1	X	X	$\overline{D}$	Justify line 4
$c =$	0				X	1							Singular cube for $G2$
$tc^5 = tc^4 \cap c =$	0	X	0	1	1	X	1	D	1	X	X	$\overline{D}$	Implications
$c =$					0	X	1						Singular cube for $G4$
$tc^6 = tc^5 \cap c =$													Inconsistency in intersection
$tc^5 =$	0	X	0	1	1	X	1	D	1	X	X	$\overline{D}$	Backtrack
$c =$					X	0	1						Singular cube for $G4$
$tc^6 = tc^5 \cap c =$	0	X	0	1	1	0	1	D	1	X	X	$\overline{D}$	Justify line 7
$c =$	1		1		0								Singular cube for $G3$
$tc^7 = tc^6 \cap c =$	0	1	0	1	1	0	1	D	1	X	X	$\overline{D}$	Justify line 7
													Done

Figure 1.4: The D-algorithm illustrated.

results in the assignments $a = 0, b = 1$, and global consistency is seen to be maintained.

The algorithm terminates in success, and provides a test pattern for the fault in question, if an error signal is propagated to some primary output, and if line justification terminates in success, i.e., assignments to all lines can be justified without any consistency problems. Thus, for our example, the algorithm terminates in success, the generated test pattern being $a = 0, b = 1$, *carry-in* $= 0$.

If the algorithm fails to propagate an error signal to a primary output due to inconsistencies in local assignments, or if the global consistency check performed in the line justification procedure fails, then the D-algorithm backtracks to the last test cube tc^j where an alternative untried choice exists. If in the process of backtracking, the test cube becomes all X's, then an alternative PDCF for the fault is selected (box 3 in Fig. 1.3a). If no more PDCFs exist, then the fault in question is declared undetectable.

The PODEM Algorithm. PODEM (Path Oriented DEcision Making) [Goe81], unlike the D-algorithm, directly searches the set of possible input patterns to obtain a test for a given fault. Hence, assignments in PODEM are made only to the primary input lines, and the values on all internal lines are computed by forward implications. This eliminates any possibility of inconsistency among values assigned to lines in the circuit which, in general, reduces backtracking compared to the D-algorithm. However, PODEM does use some information about the paths present in a gate-level model of the circuit to shorten its search. Details of the algorithm, as originally proposed by Goel in 1981, are shown in Fig. 1.5. As seen from the figure, PODEM works mostly with "objectives" which represent desired values on various internal lines, rather than actual assignments to internal lines of the circuit being considered; such assignments are made only to primary input (PI) lines (box 3 in Fig. 1.5).

PODEM starts with an input assignment A_0 in which every input line is set to X, i.e., is uninitialized, and finds successive partial input assignments $A_1, A_2, \cdots$ which maximize the possibility of an error signal generated by a fault f being propagated to a primary output. As seen from Fig. 1.5, assignments of 0 or 1 are first made to uninitialized inputs lines that are judged to have the maximum potential of generating an error signal on the faulty line. The choice of 0 or 1 is determined by a "backtrace" procedure (Fig. 1.5c) which examines the nature of the gates on the path connecting the input line under consideration to the faulty line, and is made to maximize the probability of generating an error signal (D or $\overline{D}$) on the faulty line. For example, if the faulty line is the output stuck-at-0 in a NAND gate which is fed by an uninitialized primary input signal a, then a is set to 0, leading to an error signal D on the faulty line. This is further illustrated in Fig. 1.6 using the circuit and fault from Fig. 1.2. In this case, the backtrace procedure ends with the conclusion that a 0 should be assigned to primary input *carry-in*, which results in A_1 being $a = $ X, $b = $ X, and *carry-in* $= 0$.

Every time a previously uninitialized input line is assigned 0 or 1, the logic values implied on the lines in the circuit by the current partial input assignment are computed. The result of this implication process for the circuit of Fig. 1.2, with partial input assignment A_1 is also shown in Fig. 1.6. PODEM next decides if the current partial input assignment has the potential to become a test pattern for the given fault and, if so, repeats the whole process, assigning a signal value to some other uninitialized input line. A partial assignment A_i has the potential to become a test pattern if either the faulty line

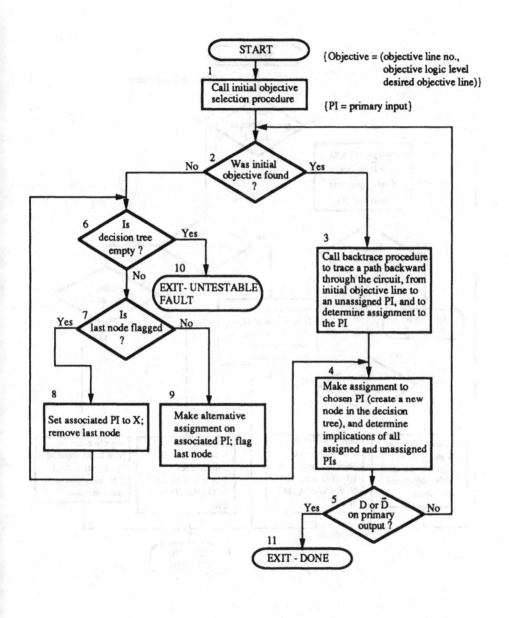

(a)

Figure 1.5: The PODEM algorithm: (a) main procedure; (b) initial objective selection procedure; (c) backtrace procedure.

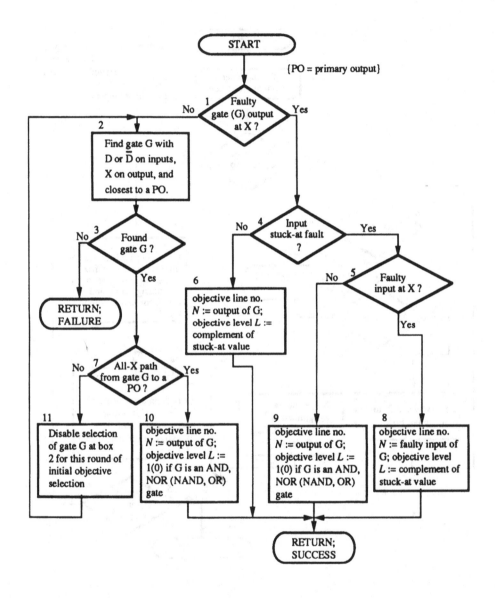

(b)

Figure 1.5: (Contd.)

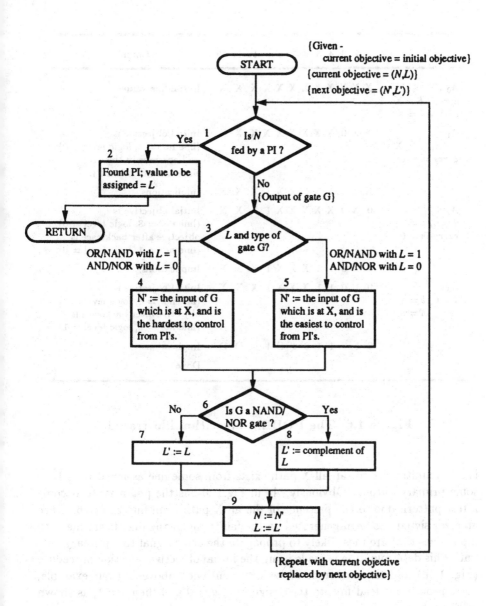

(c)

Figure 1.5: (Contd.)

	1 2 3 4 5 6 7 8 9 10 11 12	Comments
A_0 : $a = X$, $b = X$ $carry\text{-}in = X$	X X X X X X X X X X X X	Initial line values
A_1 : $a = X$, $b = X$ $carry\text{-}in = 0$	X X 0 X X X X X X X X X	Initial objective is (line no. = 3, logic level = 0); no backtrace in this case since line no. 3 is a PI.
	X X 0 X X X X X 1 X X X	Implications
A_2 : $a = 0$, $b = X$ $carry\text{-}in = 0$	0 X 0 X X X X X 1 X X X	Initial objective is (line no. = 8, logic level = 0); objective after backtrace is (line no. = 1, logic level = 0).
	0 X 0 1 1 X X X 1 X X X	Implications
A_3 : $a = 0$, $b = 1$ $carry\text{-}in = 0$	0 1 0 1 1 X X X 1 X X X	Initial objective is (line no. = 8, logic level = 0); objective after backtrace is (line no. = 2, logic level = 1).
	0 1 0 1 1 0 1 D 1 $\overline{D}$ D $\overline{D}$	Implications Done

Figure 1.6: The PODEM algorithm illustrated.

is still uninitialized, or an all-X path exists from some line assigned D or $\overline{D}$ to some primary output. Obviously, A_1 in Fig. 1.6 has the potential to become a test pattern due to the presence of the all-X path from line 22 to 26. Once an error signal has been generated on the faulty line, assignments are made to input lines that are most likely to propagate the error signal to a primary output. This determination is made using the initial objective selection procedure (Fig. 1.5b) and the backtrace procedure mentioned above. In our example, these procedures lead first to the assignment $a = 0$ and then $b = 1$, as shown in Fig. 1.6.

PODEM succeeds in finding a test input pattern if, in the process of evaluating some partial input assignment A_m, it finds that D or $\overline{D}$ signal has been implied on a primary output line. Thus, in the current example, PODEM terminates in success, since error signal $\overline{D}$ has been implied on line 26 by input assignment A_3. If, on the other hand, the algorithm finds that a current partial input assignment A_i cannot possibly detect the fault in question, it tries an alternative assignment to the last assigned input line in A_i. If none of the alternative assignments to that line works, then PODEM backtracks to the partial input assignment A_{i-1}, and tries an alternative assignment for the last assigned input line in A_{i-1}. It stops in failure if it backtracks to A_0, i.e., if it backtracks to a point where every input line becomes X. PODEM uses the same set of signal values as the D-algorithm, which limits the number of alternative assignments to each primary input line to two, viz., 0 and 1.

1.2.2 Test Generation for Sequential Circuits

Test generation is usually considerably more complex for sequential circuits than for combinational circuits due to the presence of feedback loops, as well as memory elements which are not directly observable or controllable. The combinational test generation algorithms described above require a loop-free circuit model to operate properly, and hence cannot be applied directly to sequential circuits. A test for a fault in a sequential circuit consists of a sequence of inputs rather than a single input combination. Furthermore, the response of a sequential circuit to a given sequence of inputs is a function of its initial state; hence a test also needs to initialize the circuit to a known initial state before trying to expose the fault in question.

Only a small number of useful algorithmic test generation procedures have been reported to date for sequential circuits; they are essentially modified and extended versions of the combinational test generation algorithms. For example, Kubo [Kub68], and Putzolu and Roth [Put71] have developed test generation techniques based on the D-algorithm. These procedures use a loop-free iterative array model of a sequential circuit which maps computations in consecutive time frames onto computations in consecutive cells of the array [Bre76,Fuj86]. Each cell in this model is hence identical to the sequential circuit with its feedback loops broken, and its memory elements or flip-flops (FF's) replaced by pseudo-FF's [Fuj86], which are combinational elements modeling the behavior of a FF for any single instance of time.

This modeling approach, is illustrated in Fig. 1.7 for the Huffman model of a synchronous sequential circuit. Since the circuit is synchronous, all its FFs are controlled by a common clock signal. The original sequential circuit, and the equivalent loop-free iterative array model are shown in Figs. 1.7a and 1.7b respectively. The iterative model consists of n copies of the modified loop-free circuit connected in series as shown. Each state of the iterative circuit represents the behavior of the original circuit is some specific time frame $t = i$. The dotted lines indicate that the state information for the pseudo-FF's is simply copied from the previous stage during test generation, even though there is no explicit connection between the pseudo-FF's in different stages. A typical sequential test generation algorithm for synchronous circuits is shown in Fig. 1.8. The combinational test generation subalgorithm (box 3 in Fig. 1.8)

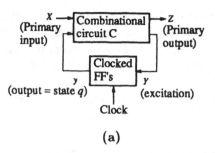

(a)

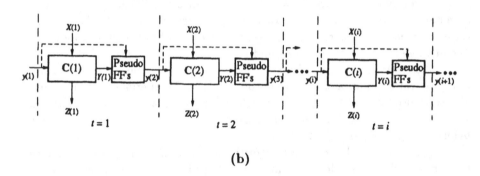

(b)

Figure 1.7: (a) Huffman model of a synchronous sequential circuit; (b) corresponding loop-free iterative model.

is left unspecified since any appropriate method can be used as long as it is suitably modified to handle the multiple faults that can arise in time frames 2 through n.

Note that that even though combinational algorithms like the D-algorithm guarantee the generation of a test, if one exists, that is not the case for the procedure of Fig. 1.8. It may fail to generate a test for a sequential circuit due to its inability to find a sequence to take the circuit to a well-defined state from the undefined initial state $y = (X, X, \cdots, X)$ (box 1 in Fig. 1.8), even though such a sequence exists [Mut76]. Moreover, practical implementations of this procedure typically use a relatively small value for MAX_I, the bound for iteration counter I, while the theoretical bound for I is exponential in the number of memory elements in the circuit, and hence, very large. As a result, the search for a test sequence may be prematurely terminated, leading to further loss of coverage. The test generation procedure of Fig. 1.8 can be extended to handle asynchronous circuits, i.e., circuits whose memory elements are not controlled by a common clock signal [Put71]. However, such extensions

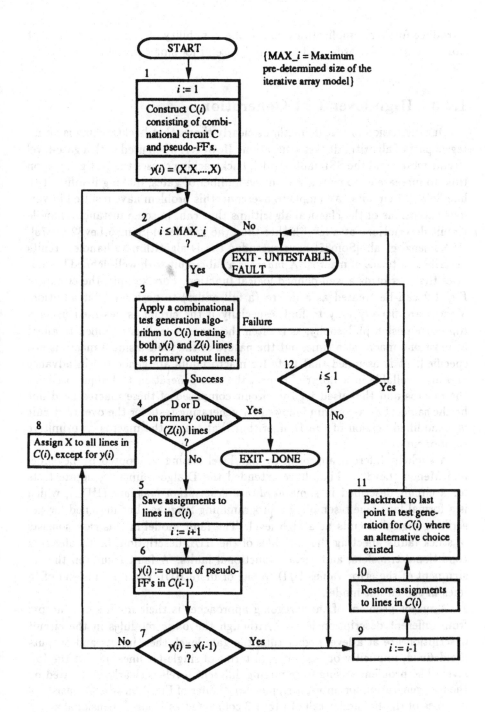

Figure 1.8: Test generation for a synchronous sequential circuit.

introduce further complications due to the possibility of generating tests that cause races and hazards, and are mainly useful for small circuits.

1.2.3   High-level Test Generation

Our discussion of testing methods clearly shows that backtracking is an integral part of algorithmic test generation. However, when used with a gate-level circuit model and the SSL fault model, backtracking causes the test generation time to increase exponentially with the number of gates, making it difficult to handle VLSI circuits. Attempts to overcome this problem have resulted in various extensions of the classical algorithms that can, in some instances, handle circuit descriptions at a level higher than the gate level [Som85,Lev82,Lee76].

Somenzi et al. [Som85] have extended the D-algorithm to handle circuits described in terms of macros. A *macro* is a subcircuit with well-defined boundaries that performs a well-defined logical function. For example, the circuit of Fig. 1.2 can be treated as a macro (a full adder) to perform 1-bit addition. Macros are usually easy to find, especially if the circuit is designed using a top-down design philosophy, as is often the case. A fault is assumed to affect at most one macro at a time, but the nature of the fault inside a macro is not specified. It is assumed that tests for macro faults are generated in advance and are available in a library. Hence, the test generation technique used for the macros and that used for the circuit composed of these macros need not be the same. The algorithm followed for generating tests for the overall circuit is a modified version of the D-algorithm that treats the macros as primitive components.

A slightly different approach to high-level testing is adopted by Levendel and Menon [Lev82]. They have extended the D-algorithm to generate tests for a circuit described in some *hardware description language* (HDL), which is a formal language resembling a programming language but designed for describing digital circuits at a high level. The fault model in this case consists of stuck faults affecting the variables of the HDL description, faults affecting conditional transfers, and certain functional faults. A fault results in the assignment of the error values $D/\overline{D}$ to one or more HDL variables instead of to lines in the circuit model.

A major problem of the foregoing approaches is their mixing of concepts from different description levels. Although the basic modules in the circuit description are at a level higher than the gate level, their inputs and outputs are defined, implicitly or explicitly, in terms of single-bit lines, i.e., at the low level. The problem arising from mixing different levels is clearly illustrated by the test generation for an n-bit ripple-carry adder of Fig. 1.9a which consists of n copies of the full adder cell of Fig. 1.2 connected as a one-dimensional array. Obviously, the number of components in this array is proportional to n, the size of the array. Hence, the complexity of gate-level test generation for this circuit is a function of n, which usually grows exponentially with n. In order to use the approach of [Som85], we first need to identify the macros in this circuit; an obvious assumption is that each full adder cell constitutes a macro; see Fig. 1.9b. In this case, the macro size remains constant for all adder sizes, but the number of macros in the adder design is obviously n. Hence, the complexity of test generation using the macro-based model is still an exponential function in n. On the other hand, we could also treat the whole adder array as a macro, since it performs a well-defined logical function. However, this macro has $2n+1$ input lines and $n+1$ output lines, implying that the number of possible input patterns to the macro increases exponentially with n. Thus, in both cases, the reduction in the number of components from working at a higher level than the gate level is offset to a large extent by the increased complexity of representing the input/output behavior of the macro. Somenzi et al. have also introduced some heuristic observability and controllability measures to alleviate the problem of representing the behavior of complex macros, but the accuracy of such measures is difficult to estimate.

A similar problem occurs in the work of Levendel and Menon [Lev82]. For example, in their approach, the adder of Fig. 1.3 would be represented by a primitive HDL function with $2n+1$ input variables, and $n+1$ output variables. Manually constructed tables of useful input patterns that propagate error signals through such common high-level primitive blocks as adders, counters, etc., are given in [Lev82]. However, these tables appear to have been constructed in an ad hoc manner, and no general construction technique for such tables is defined. Furthermore, lack of experimental data makes it difficult to estimate the relative gains made by the proposed technique over the conventional D-algorithm.

Lee has presented a different approach to high-level test generation which uses a circuit model consisting of word gates only [Lee76]. This model, called the *vector canonical realization*, is a generalization of the canonical sum-of-minterms realization of switching functions, in which the circuit behavior is described in terms of multibit input/output vectors or words, and a canonical expression is formulated for each output vector, analogous to the formulation of the canonical expressions for switching functions. However, such a realization often contains a vast number of redundant gates, and is very inefficient. For example, the vector canonical realization of a 2-bit adder contains 51 word gates, each of which is made up of two gates, compared to 26 gates required for a traditional canonical realization, and 18 gates required for an implementation using the structure of Fig. 1.3. Hence, tests generated for vector canonical realizations appear inappropriate for practical circuits.

Over the last decade, various other high-level test generation techniques have been proposed that differ significantly from the classical approaches. An important class of such techniques [Fri73,Sri81a,Sri81b,Che86] construct

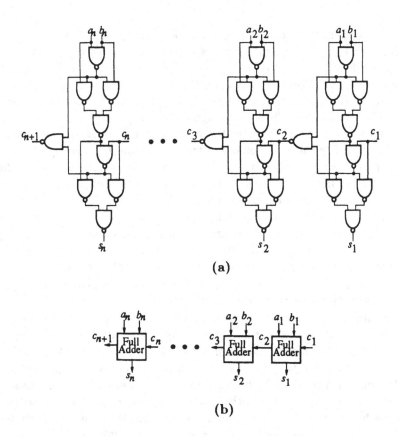

(a)

(b)

Figure 1.9: Models of a ripple-carry adder: (a) gate level; (b) register level.

pseudo-exhaustive test sets for array circuits like that of Fig. 1.3 using only a high-level description of the input/output behavior of the modules in the array. Another test generation method, proposed in [Tha78,Tha80], requires a register-level description in terms of the instruction behavior of the circuit under test. A common limitation of these techniques is their lack of generality. For example, the pseudo-exhaustive technique has been found useful for array-type circuits, but is difficult to extend to general circuits. The approach of [Tha80], on the other hand, is applicable only to instruction-set processors. Moreover, all such high-level methods are incompatible with standard gate-level techniques in that they cannot be reduced to the traditional methods merely by changing the level of the circuit and the fault models used. This means that their effectiveness compared to conventional methods is difficult or impossible to determine.

In summary, conventional test generation algorithms like the D-algorithm and PODEM provide generality, but suffer from computational inefficiency due to the low level of the circuit and fault models they employ. Various high-level test generation algorithms have been proposed using circuit models consisting of modules whose internal structure may or may not be known. Other high-level test generation techniques have been proposed that use circuit descriptions in terms of HDL constructs. Although they are useful for special classes of circuits, this usefulness is often limited by the mixing of concepts from various levels or by the lack of generality of the circuit or fault models used.

1.2.4   Fault Simulation

As mentioned in Section 1.1, fault simulation is usually an important part of the test generation process for logic circuits. It is used for various purposes during and after test generation, such as determining the coverage achieved by a set of tests, analyzing behavior in the presence of faults, etc. It is often used concurrently with the test generation process to identify all faults detected by a generated test. These faults can be dropped from the fault list for subsequent test generation cycles, thus avoiding the generation of too many tests for any one fault in the circuit.

The typical input to a fault simulator consists of a circuit model, a list of input patterns, and a list of faults applicable to the circuit. Usually, the SSL fault model is used, implying only one faulty line in the circuit, although multiple faults can be handled without substantial difficulty. The simulator analyzes the behavior of the circuit when each of the input patterns is applied in the presence of faults from the fault list; one pass through the whole circuit for a given input pattern is referred to as a *simulation cycle*. The output of the simulator usually consists of information about logic values as well as timing of signals appearing on user-specified groups of lines, at the end of each simulation cycle.

Based on the circuit representation technique used, fault simulators can be either *compiled*, i.e., the circuit is implicitly described in the form of (executable) compiled code, or *table-driven*, i.e., the circuit is described using tables [Fuj86,Bre76]. In the table-driven case, the simulator is said to be *event (activity) directed* if only the active portion of the circuit is simulated in each simulation cycle, [Fuj86,Bre76]. Fault simulators are also classified by the number of faults and input patterns processed simultaneously in any simulation cycle. The simplest type of fault simulators in this regard are the single-pattern single-fault simulators which process one input pattern and one fault per simulation cycle. However, at the gate level, most fault simulators process many faults simultaneously, and are classified into three broad kinds:

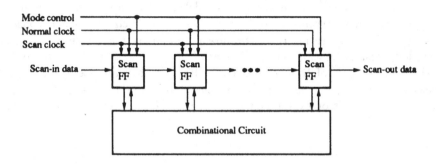

Figure 1.10: General structure of a sequential circuit with scan design.

parallel, deductive, and concurrent fault simulators. Detailed descriptions of these three types of fault simulators are available in [Bre76,Fuj86].

Most research in fault simulation using higher-than-gate-level circuit descriptions has focussed on using functional or behavioral models of circuit components [Cha88,Gho88,Hir81,Rog87]. A notable example of this approach is found in the hierarchical fault simulator implemented by Rogers in [Rog87], which works with functional models of components at the higher levels, the functional models of the high-level components in a circuit being constructed automatically from their gate-level models. Later in this book, we will present a a simple table-driven activity-directed single-pattern single-fault simulator which can handle our high-level circuit and fault models.

1.2.5   Design for Testability

Over the last two decades, it has become increasingly clear that newer and better test generation techniques alone cannot provide the complete answer to the test generation problem for large circuits. This is mainly due to the inherently intractable nature of the problem; the test generation problem is provably NP-complete for combinational circuits [Iba75]. Hence, large circuits need to be designed to be easily testable. However, reductions in testing complexity are not free, and a penalty is usually paid in extra logic or hardware overhead that must be added to enhance testability of the circuit in question. Moreover, most DFT techniques consist of ad hoc modifications to existing circuits or components, and are often not practical due to their excessive over-

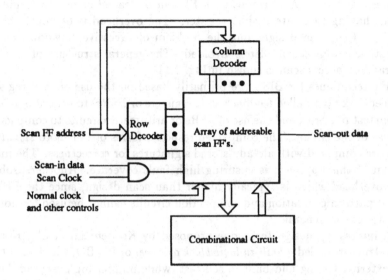

Figure 1.11: General structure of a sequential circuit with random-access scan.

head. The systematic DFT techniques that have been proposed can be broadly classified into two categories: DFT to facilitate testing with external ATE, and DFT to allow built-in self-test (BIST). Some representative techniques in both categories are discussed next.

DFT techniques to facilitate external testing attempt to reduce the complexity of test generation for sequential circuits by allowing access to internal lines and memory elements. In order to minimize the circuit overhead, most of these techniques link in a chain some or all of the flip-flops (FFs) in the circuit, thus forming a shift register. This temporary shift register can be loaded with any desired pattern, or its contents can be read during the testing process. Such designs are referred to as *scan designs* [Fuj86], and the shift register is called a *scan register*. The most notable scan technique is level-sensitive scan design (LSSD) [Eic78], which also eliminates race and hazard-related problems in sequential circuits.

The general structure of a scan design is shown in Fig. 1.10. Large scan designs suffer from the excessive lengths of their scan registers, which can slow down application of test patterns considerably. They also incur some hardware overhead due to the fact that the scanable FFs tend to be larger than ordinary FFs. Ando [And80] has suggested an interesting variation of the scan approach which allows all the FFs in a circuit to be addressed individually using extra address lines that have no use for normal operation of the circuit; this is called *random-access scan*. As a result, each FF can be loaded or read individually, without having to create a shift register. The overhead is obviously higher than the other scan design, but the problem of excessive slowdown of test application is almost completely avoided. The general structure of a circuit with random-access scan is shown in Fig. 1.11.

DFT techniques for BIST are primarily based on the use of shift registers with feedback (also called feedback shift registers or FSRs) to generate pseudo-random test patterns, and the use of FSRs or arithmetic circuits to compact the response of the circuit. The compacted response forms one or more *signatures* which are compared with already stored signatures for correctness. The major problem of this approach is ensuring high fault coverage. Another problem is the overhead which is typically higher than scan design, since the FF's in the test pattern generation and compaction circuits cannot be used for normal computation, in general.

An interesting design technique proposed by Koenemann et al. [Koe79], uses a structure called *a built-in logic block observer* or BILBO, which facilitates both external testing and built-in self test, while minimizing overhead due to wasted FFs. A BILBO is a register with two extra control inputs B_1 and B_2, and some extra logic placed between successive FFs such that the BILBO can be configured in any of three modes: (a) parallel-in parallel-out register, (b) shift register, (c) feedback shift register. The structure of a BILBO register is shown in Fig. 1.12, and a typical application is illustrated in Fig. 1.13. In this case, the three modes of operation mentioned above correspond to the control settings $B_1B_2 = 10$, $B_1B_2 = 01$, and $B_1B_2 = 11$ respectively. The combination of control inputs ($B_1B_2 = 00$) can be used to reset the entire register. Obviously, the FFs in a BILBO can be used as regular FFs during normal operation. On the other hand, the FSR configuration can be utilized to test the blocks $C_1, \cdots, C_n$ using the BIST technique. Finally, the shift register configuration can be used to scan tests and responses in and out of the registers, so BILBO also facilitates external testing using ATE.

Finally, we consider a circuit partitioning technique proposed by Bozorgui-Nesbat and McCluskey [McC80] which helps in reducing the complexity of both external testing and built-in self test. In this approach, the circuit under test is partitioned into well-defined blocks (similar to the macros considered earlier) that have disjoint sets of inputs, outputs, and internal linking buses; see Fig. 1.14. Multiplexers (MUX's) and extra interconnections are now added to the existing blocks such that in the test mode, each block can have test patterns applied to it directly from the primary inputs, and its response can be observed at primary outputs. The general structure of the modified circuit, and its configurations in the normal and test modes are also shown in Fig. 1.14.

Thus, we see that the majority of systematic DFT techniques try to reduce the complexity of sequential circuit testing either by providing relatively easy

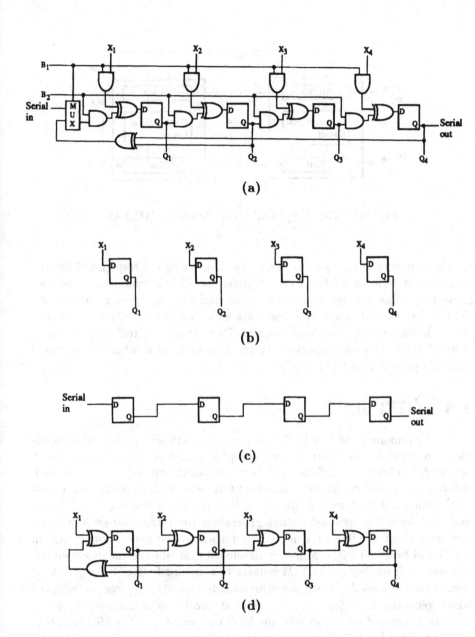

Figure 1.12: Structure and configurations of a typical 4-bit BILBO register: (a) logic diagram; (b) parallel register mode ($B_1B_2 = 10$); (c) shift register mode ($B_1B_2 = 01$); (d) feedback shift register mode ($B_1B_2 = 11$).

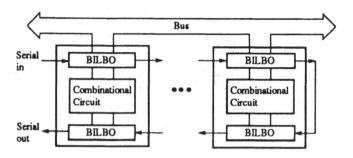

Figure 1.13: A typical application of BILBO.

access to internal memory elements, or by compacting the response of the circuit under test using feedback shift registers or other mechanisms. However, these techniques usually neglect the combinational logic present in circuits. DFT techniques addressing this issue are either ad hoc in nature, or are impractical due to large overhead [Fuj86]. Thus, there is a real need for more efficient DFT methods, especially in the case of large combinational blocks typically present in a VLSI circuit.

1.3   OUTLINE

In the remainder of this book, we present an efficient high-level methodology to improve test generation for complex digital circuits. Our approach uses high-level circuit and fault models, minimizes the mixing of concepts from different levels, and reduces to a standard test generation algorithm when low-level circuit and fault models are used. We introduce a new high-level circuit and fault modeling technique which generalizes the traditional gate-level circuit model and the SSL fault model. A high-level test generation algorithm VPODEM based on PODEM is then developed that employs the proposed circuit and fault models. VPODEM reduces to standard PODEM if a gate-level circuit model is used. Thus, a hierarchical testing strategy can be adopted which generates tests for general circuits at two levels of description. First, tests are generated for a high-level model of the circuit, and the SSL fault coverage provided by these tests is determined. Next, tests are generated for the remaining undetected SSL faults in the gate-level model of the circuit using the same algorithm, so 100 percent SSL fault coverage can always be obtained. Finally, we present some systematic DFT techniques which significantly simplify the testing complexity of large and important combinational circuits such as fast ALU's, with reasonably small hardware overhead.

Chapter 2 describes our high-level circuit and fault modeling technique.

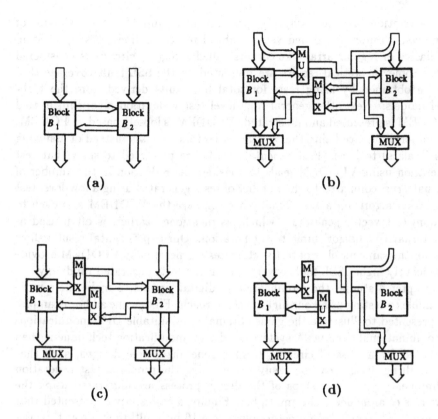

Figure 1.14: Partitioning approach of Bozorgui-Nesbat and Mc-Cluskey: (a) unmodified circuit; (b) modified circuit; (c) normal mode of operation; (d) test mode, testing block B_1.

First, a vector notation is introduced which describes component behavior hierarchically in terms of matrices called vector sequences. This provides a convenient tool for describing the test vectors to be generated later. The proposed high-level circuit and fault models are presented next. The high-level circuit model employs register-level components like word gates, multiplexers, and adders, which are interconnected by multibit buses. The basic fault model consists of bus faults, in which lines in a bus may be stuck at 0 or 1. In order to make the problem manageable, we restrict attention to an important subset of bus faults designated total bus faults. The proposed modeling techniques are then illustrated with a case study for a class of useful array-like circuits called k-regular circuits.

Chapter 3 presents our hierarchical test generation scheme. It starts with the description of a new way of representing the input/output behavior of high-level components, called vector cubical representation. Test generation for classes of regular array-like circuits called k-regular circuits is considered next. Some theoretical results are presented on the SSL fault coverage that can be obtained by using tests for total bus faults derived from the high-level models. Next, the general high-level test generation algorithm named VPODEM is presented and illustrated. VPODEM is loosely based on PODEM, the major difference being that it assigns vectors to buses instead of assigning bits (scalars) to lines. Experimental results are presented to show that test generation using VPODEM leads to considerable reduction in the number of test patterns, compared to the number of tests generated using a low-level test generation algorithm alone. Finally, we compare the VPODEM approach to random test vector generation which, as mentioned earlier, is often used as an alternative to algorithmic test generation. Our experimental results show that for the same number of test vectors, tests generated by VPODEM provide significantly higher SSL fault coverage than the tests generated randomly.

Chapter 4 studies the modification of digital circuits to make them more amenable to testing using our high-level approach. First, some ad hoc examples are presented to illustrate the gains obtainable by suitable DFT modifications to combinational circuits. A systematic design modification technique is then presented for a class of circuits, termed generalized tree designs, such that the resulting designs can be easily tested using the proposed test generation technique. The various steps of the design process are illustrated using the examples of an adder and a multiplier. Finally, a case study is presented that applies the proposed design methodology to a 16-bit multi-function arithmetic-logic unit.

Chapter 5 summarizes the high-level test generation technique described in this book, and suggests some possible extensions of our approach. Proofs of the theorems are provided in an appendix.

Chapter 2

CIRCUIT AND FAULT MODELING

This chapter begins with a brief introduction of the vector sequence (VS) notation which will be used in this book as a tool for describing the behavior of high-level components, as well as for representing the faults and test patterns generated at different levels of abstraction. The general high-level circuit and fault modeling techniques are then presented. The chapter concludes with a detailed study of model construction at different abstraction levels for the special case of k-regular circuits. This illustrates the advantages of our approach over conventional modeling techniques.

2.1 VECTOR SEQUENCE NOTATION

The VS notation first proposed in [Hay80] and subsequently expanded by us in [Bha85], has $n \times m$ arrays called *vector sequences* (VS's), where n and m are dynamic parameters, as the fundamental information units. This notation is primarily intended to represent input/output behavior of circuits at different levels of complexity. For example, the behavior of the gate-level circuit of Fig. 2.1a can be written in the VS form as

$$
\begin{bmatrix}
0 & 0 & 0 & 0 & 1 & 1 & 1 & 1 \\
0 & 0 & 1 & 1 & 0 & 0 & 1 & 1 \\
0 & 1 & 0 & 1 & 0 & 1 & 0 & 1
\end{bmatrix}
\Big/
\begin{bmatrix}
0 & 0 & 0 & 0 & 1 & 0 & 0 & 0 \\
0 & 0 & 0 & 0 & 0 & 1 & 0 & 0 \\
0 & 0 & 0 & 0 & 0 & 0 & 1 & 0 \\
0 & 0 & 0 & 0 & 0 & 0 & 0 & 1
\end{bmatrix}
\tag{2.1}
$$

The same circuit can be represented at a higher level as shown in Fig. 2.1b, with the following VS's defining its input/output behavior:

$$
\begin{bmatrix}
V_1 & V_3 \\
V_2 & V_2
\end{bmatrix}
\Big/
\begin{bmatrix}
V_4 & V_5
\end{bmatrix}
\tag{2.2}
$$

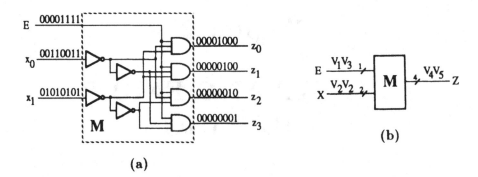

Figure 2.1: Two-to-four decoder: (a) gate-level model; (b) high-level model.

where

$$V_1 = [\ 0\ \ 0\ \ 0\ \ 0\ ], \quad V_2 = \begin{bmatrix} 0 & 0 & 1 & 1 \\ 0 & 1 & 0 & 1 \end{bmatrix}, \quad V_3 = [\ 1\ \ 1\ \ 1\ \ 1\ ],$$

$$V_4 = \begin{bmatrix} 0 & 0 & 0 & 0 \\ 0 & 0 & 0 & 0 \\ 0 & 0 & 0 & 0 \\ 0 & 0 & 0 & 0 \end{bmatrix}, \quad V_5 = \begin{bmatrix} 1 & 0 & 0 & 0 \\ 0 & 1 & 0 & 0 \\ 0 & 0 & 1 & 0 \\ 0 & 0 & 0 & 1 \end{bmatrix} \tag{2.3}$$

We will see in this chapter that VS's provide a compact way of representing the input/output behavior of components in a circuit. Various VS operators, to be described shortly, can be used to derive the input/output behavior of components described at a higher level of abstraction, given the input/output behavior of the lower-level components in the VS notation. This allows us to express concisely the test patterns for basic building blocks of digital circuits like gates, multiplexers, etc., using the VS notation. Moreover, the cubical representation of behavior of components [Rot66] described earlier, which is employed by many test generation algorithms, can be expanded to obtain a vector cubical representation of the behavior of high-level components. This facilitates the development of a vector-oriented test generation algorithm presented later in Chapter 3.

In general, the elements of a VS are either general VS's or are primitive VS's called *atoms*, meaning they consist of indivisible symbols like 0 and 1. Two VS's are said to be *space (time) compatible* if they have the same number of rows and columns. All VS's of interest can be built from atoms using two basic VS operators, viz., external time expansion and external space expansion, defined below [Hay80]:

(1) *External Time Expansion*: This operator concatenates two space-compatible VS's $[S]$ and $[T]$ in the time dimension thus:

$$[S] \cdot [T] = [S \quad T]$$

(2) *External Space Expansion*: This operator concatenates two time-compatible VS's $[S]$ and $[T]$ in the space dimension thus:

$$[S] \odot [T] = \begin{bmatrix} S \\ T \end{bmatrix}$$

The procedure for constructing larger VS's from smaller ones is defined recursively as follows: (i) If a is an atom then $[a]$ is a VS. (ii) If V_1 and V_2 are two space-compatible VS's then $V_1.V_2$ is a VS. (iii) If V_1 and V_2 are two time-compatible VS's then $V_1 \odot V_2$ is a VS. (iv) Nothing else is a VS.

Four more operators, viz., $\times, \otimes, S_\alpha$ and P_α, have also been defined on VS's, which provide the ability to specify complex expansion and contraction operations concisely. The $\times$ and $\otimes$ operators were defined in [Hay80], while S_α and P_α were introduced by us in [Bha85]. Their definitions are as follows:

(3) *Internal Time Expansion*: If S and T are two time- and space-compatible $n \times m$ VS's of the form

$$S = \begin{bmatrix} S_{11} & S_{12} & \cdots & S_{1m} \\ S_{21} & S_{22} & \cdots & S_{2m} \\ \cdot & \cdot & & \cdot \\ \cdot & \cdot & & \cdot \\ S_{n-1,1} & S_{n-1,2} & \cdots & S_{n-1,m} \end{bmatrix}$$

$$T = \begin{bmatrix} T_{11} & T_{12} & \cdots & T_{1m} \\ T_{21} & T_{22} & \cdots & T_{2m} \\ \cdot & \cdot & & \cdot \\ \cdot & \cdot & & \cdot \\ T_{n-1,1} & T_{n-1,2} & \cdots & T_{n-1,m} \end{bmatrix} \quad (2.4)$$

then their internal time expansion $S \times T$ is an $n \times m$ VS whose (i,j)th element is $S_{ij}.T_{ij}$.

(4) *Internal Space Expansion*: If S and T are two $n \times m$ VS's as defined in (2.4), then their internal space expansion $S \otimes T$ is an $n \times m$ VS whose (i,j)th element is $S_{ij} \odot T_{ij}$.

(5)   *Select*: The select operator S_α acting on a VS V produces a VS $S_\alpha(V)$ consisting of the rows of V specified by α. If α denotes a row index range $a..b$ $(a \leq b)$, then $S_\alpha(V)$ is composed of the rows a through b of V. If α denotes a row index set $(a_1, a_2, \cdots, a_n)$, then $S_\alpha(V)$ consists of the (not necessarily distinct) rows $a_1, a_2, \cdots, a_n$ of V. If $a > b$, or if V does not contain any of the rows specified by α, then $S_\alpha(V)$ is taken to be the null vector ϕ.

(6)   *Project*: The project operator P_α acting on a VS V produces a VS $P_\alpha(V)$ consisting of the columns of V specified by α. If α denotes a column index range $a..b$ $(a \leq b)$, then $P_\alpha(V)$ is composed of the columns a through b of V. If α denotes a column index set $(a_1, a_2, \cdots, a_n)$, then $P_\alpha(V)$ consists of the (not necessarily distinct) columns $a_1, a_2, \cdots, a_n$ of V. If $a > b$, or if V does not contain any of the columns specified by α, then $P_\alpha(V)$ is taken to be the null vector ϕ.

For example, if $S = T = \begin{bmatrix} 1 & 0 \\ 0 & 1 \end{bmatrix}$ then

$$S \cdot T = \begin{bmatrix} 1 & 0 & 1 & 0 \\ 0 & 1 & 0 & 1 \end{bmatrix}, \qquad S \odot T = \begin{bmatrix} 1 & 0 \\ 0 & 1 \\ 1 & 0 \\ 0 & 1 \end{bmatrix}$$

$$S \times T = \begin{bmatrix} 1 & 1 & 0 & 0 \\ 0 & 0 & 1 & 1 \end{bmatrix}, \qquad S \otimes T = \begin{bmatrix} 1 & 0 \\ 1 & 0 \\ 0 & 1 \\ 0 & 1 \end{bmatrix}$$

$$S_{(1)}(S) = \begin{bmatrix} 1 & 0 \end{bmatrix}, \qquad P_{(2)}(S) = \begin{bmatrix} 0 \\ 1 \end{bmatrix}$$

Certain "standard" vector sequences have been defined using the expansion operators alone [Hay80]. For example, 0_n represents the vector of n 0's; 1_n is the vector of n 1's; A_n is the vector of size n with alternating 1's and 0's; C_n is the $n \times 2^n$ counting sequence (which is the output of an n-bit counter), and D_n is the $n \times n$ diagonal sequence. Their recursive definitions are given in Fig. 2.2a where a prime denotes the transpose operator. The standard VS's are illustrated for $n = 4$ in Fig. 2.2b.

A VS expression of the form $V = f(S, T)$ can be interpreted as describing the behavior of a functional block whose inputs are the VS's S and T, and whose output is the VS V; see Fig. 2.3a. For example, if we select f to be the operator $\odot$, then the corresponding functional block performs a simple merger of two buses into one bus, as shown in Fig. 2.3b. The output bus in this case is a juxtaposition of the two input buses. The element of Fig. 2.3b, which is primarily an abstract notational device, is called the *merge* element and provides a basic way of representing the construction of buses at higher levels

$$0_1 = 0; \; 0_n = 0_{n-1} \circledcirc 0$$
$$1_1 = 1; \; 1_n = 1_{n-1} \circledcirc 1$$
$$A_1 = 1; \; A_{2n} = A_{n-1} \circledcirc 0; \; A_{2n+1} = A_{2n} \circledcirc 1$$
$$C_1 = 0.1; \; C_n = (0'_{2^n-1} \circledcirc C_{n-1}) \cdot (1'_{2^n-1} \circledcirc C_{n-1})$$
$$D_1 = 1; \; D_n = (D_{n-1} \circledcirc 0'_{n-1}) \cdot (0_{n-1} \circledcirc 1)$$

(a)

$$0_4 = \begin{bmatrix} 0 \\ 0 \\ 0 \\ 0 \end{bmatrix}, \; 1_4 = \begin{bmatrix} 1 \\ 1 \\ 1 \\ 1 \end{bmatrix}, \; A_4 = \begin{bmatrix} 1 \\ 0 \\ 1 \\ 0 \end{bmatrix}$$

$$C_4 = \begin{bmatrix} 0000000011111111 \\ 0000111100001111 \\ 0011001100110011 \\ 0101010101010101 \end{bmatrix}, \; D_4 = \begin{bmatrix} 1000 \\ 0100 \\ 0010 \\ 0001 \end{bmatrix}$$

(b)

Figure 2.2: (a) Recurrence relations defining standard VS's; (b) Standard VS's for $n = 4$.

of abstraction. The converse operation, viz., the splitting of a bus into two or more buses, can be performed by another class of abstract logic elements called *fanout elements*; see Fig. 2.3c. Merge and fanout elements, which are further examined in the next section, allow us to represent arbitrary changes in bus sizes in a simple manner, and play an important role in the construction of high-level models of general circuits.

The VS notation also provides a concise but flexible way of expressing the input/output behavior of components, suggesting that it may be possible to use this notation in describing tests for these components. For example, a complete test input set and the corresponding response set for an n-input NAND gate can be easily seen to be

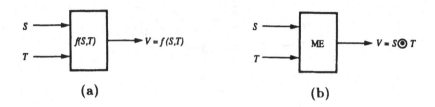

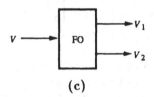

Figure 2.3: Interpretation of VS operators as functional blocks: (a) general operator element performing function f; (b) merge element; (c) fanout element.

$$\begin{bmatrix} 0 & 1 & 1 & & 1 & 1 \\ 1 & 0 & 1 & & 1 & 1 \\ 1 & 1 & 0 & & 1 & 1 \\ . & . & . & \cdots & . & . \\ . & . & . & & . & . \\ . & . & . & & . & . \\ 1 & 1 & 1 & & 0 & 1 \end{bmatrix} / \begin{bmatrix} 1 & 1 & 1 & \cdots & 1 & 0 \end{bmatrix}$$

Using standard VS's, this can be represented more compactly as

$$\overline{D}_n \cdot 1_n / 1^{\cdot n} \cdot 0$$

This NAND VS can be readily extended to the case of an m-bit NAND word gate, the test/response set being

$$\left[\overline{D}_n \cdot 1_n \right]^{\otimes m} / \left[1^{\cdot n} \cdot 0 \right]^{\otimes m}$$

Similarly, the test/response set for pin faults (SSL faults on the primary input lines only) in a 2^p-input q-bit multiplexer can be represented as follows:

$$\left[D_{2^p}^{\otimes q} \odot C_p \right] \cdot \left[\overline{D}_{2^p}^{\otimes q} \odot C_p \right] / 1_q^{\cdot 2^p} \cdot 0_q^{\cdot 2^p} \tag{2.5}$$

As we move to more general circuits, it becomes necessary to use both expansion and contraction (select and project) operators in the expressions describing the tests sets. This is illustrated by the test set for pin faults in a general m-input q-bit multiplexer. In this case, a VS expression of the form in (2.5) cannot be used directly, since, in general, $m \neq 2^p$ for an integer p. A natural choice of p in this case is $p = \lceil \log_2 m \rceil$, which implies that we need to apply only a part of the standard VS C_p to the multiplexer's control lines. The project operator allows us to specify this, yielding the following description of the multiplexer test/response set:

$$\left[D_m^{\otimes q} \odot \mathrm{P}_{(1..m)}\left(C_p\right) \right] \cdot \left[\overline{\mathrm{D}}_m^{\otimes q} \odot \mathrm{P}_{(1..m)}\left(C_p\right) \right] / 1_q^{\bullet m} \cdot 0_q^{\bullet m}$$

As shown in the next section, the select operator is particularly useful for describing in a uniform way, the complex signal changes created by the fanout (branching) of multiline buses.

The VS notation thus provides a useful descriptive tool for hierarchical test generation, one that is capable of handling circuits at various levels of abstraction. As mentioned earlier, this notation can also be used to construct a vector cubical representation of input/output behavior of high-level components, leading to a vector-oriented test generation procedure. In the rest of this chapter, and in Chapter 3, we investigate the role of such vectors in test generation algorithms where the circuit and fault models are defined at two or more levels.

2.2   CIRCUIT AND FAULT MODELS

Test generation for a digital circuit consists of identifying a set of input patterns that, applied to the circuit in the presence of some fault of interest, cause the circuit to produce an error signal (0 instead of 1, or vice versa) at some observable output. Traditional test generation techniques use a gate-level circuit model, and allow any single line in the circuit to be faulty (the single-stuck-line or SSL fault model). However, the time required to generate tests using conventional techniques increases exponentially with the number of components in the circuit, and becomes prohibitive for VLSI circuits. This motivates the development of other approaches to test generation. A few such test generation techniques, as already discussed in Chapter 1, have been proposed that model circuits in terms of purely functional blocks or in terms of fairly artificial constructs, e.g., vector canonical forms. As discussed in Chapter 1, previous high-level test generation techniques suffer from three major problems in varying degrees, viz., mixing of concepts from various descriptive levels, restriction to small classes of circuits, and lack of compatibility with traditional test generation techniques. In the rest of this chapter, we will develop

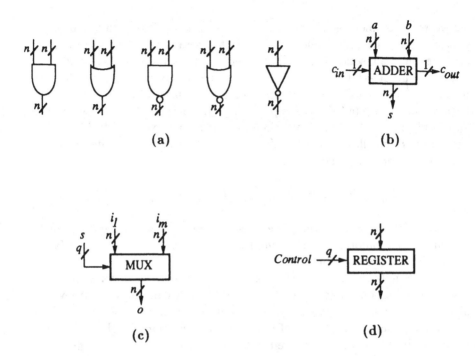

Figure 2.4: **Commonly used high-level components: (a) word gates; (b) adder; (c) multiplexer; (d) register.**

a hierarchical circuit and fault modeling approach which provides a solution to these problems.

2.2.1   Circuit Model

The proposed high-level circuit model is composed of components such as word gates, multiplexers, adders, registers, etc., connected to one another using buses of appropriate sizes. Some high-level components commonly used in our circuit model are illustrated in Fig. 2.4. Note that this level corresponds approximately to the register or functional level of circuit design. Also note that the bus sizes (denoted by n and q in Fig. 2.4) are parameters whose values vary with the circuit's level of abstraction. When the bus size is reduced to one, a word gate becomes a single gate, and a bus is the same as a line. Thus our modeling approach constitutes a natural generalization of the classical gate-level modeling technique. As we will see in the remainder of this book, this approach makes truly hierarchical modeling of circuits possible, while minimizing the mixing of levels in the circuit models, and helps in reducing the overall test generation effort.

A feature of our circuit modeling technique is its ability to combine repeated subcircuits in the gate-level model of a circuit into high-level subcircuits consisting of word gates and fanout elements only. Functional blocks like the carry-lookahead generator in the 74181 IC, which do not possess much regularity may be retained as primitive high-level components. However, the construction of a truly high-level model from a given gate-level circuit may be complicated by the fact that the generalization of single-bit lines to multibit buses introduces some unique modeling issues at the higher level. The more important of these issues are general fanout structures and bidirectional signal flow. In this section, we introduce a modeling technique for general bus fanout applicable to both high- and low-level circuits. A technique is also developed for constructing realistic high-level models of circuits containing bidirectional buses.

Fanout Element. A *fanout* element is a component with one input bus X and m output buses $Z_1, \cdots, Z_m$ such that an input vector V applied to X produces output vectors $S_{\alpha_i}(V)$ on Z_i for $1 \leq i \leq m$, where S_{α_i} is the select operator defined in Section 2.1. Examples of fanout elements and other interconnection structures commonly used by our model are presented in Fig. 2.5. A fanout element is said to be *regular* if $S_{\alpha_i}(V) = V$ for $i = 1, \cdots, n$, i.e., the input and output buses are all the same size and carry the same signal vectors. It is a generalization of the simplest type of fanout occurring in gate-level circuit models, in which all fanout elements are regular with input and output buses of size one. They are, in fact, simple wiring nets driven by a single signal source.

To illustrate the circuit modeling process, we apply it to some standard MSI circuits [Tex85]. The gate-level model $\mathbf{M}^G$ of a 4-bit 2-to-1 multiplexer is shown in Fig. 2.6a. The corresponding high-level model $\mathbf{M}^H$ appears in Fig. 2.6b. The subcircuits marked $M_{2,1}$ through $M_{2,4}$ of $\mathbf{M}^G$ are represented in $\mathbf{M}^H$ as a word-oriented subcircuit $M_2{}^4$. As shown in Fig. 2.6b, several fanout elements also need to be introduced in $\mathbf{M}^H$ that change the size of the control buses from one to four. As a second example, consider the standard 74181 ALU/function generator [Tex85], whose $\mathbf{M}^G$ and $\mathbf{M}^H$ are shown in Fig. 2.7. The subcircuits M_1 through M_4 of $\mathbf{M}^G$ are again grouped into a word-oriented high-level subcircuit M^4 in $\mathbf{M}^H$. The subcircuit M_5 of $\mathbf{M}^G$ does not possess much regularity in its structure, but is easily seen to be a modified carry-lookahead adder that can also function as a XNOR word gate depending on the control signal $\bar{M}$. Hence, this portion of the circuit is modeled as a primitive component MCLA with four input buses, two of size four and two of size one, and four output buses, as shown in Fig. 2.7b. $\mathbf{M}^H$ for the 74181 ALU/function generator is completed by connecting M^4 and MCLA via two four-bit buses $B1$ and $B2$ in Fig. 2.7b.

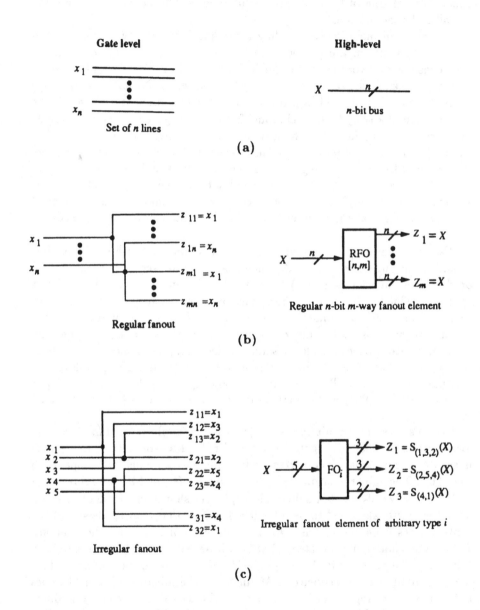

Figure 2.5: Representation of interconnection structures in gate-level and high-level circuit models; (a) simple interconnection; (b) regular fanout; (c) irregular fanout.

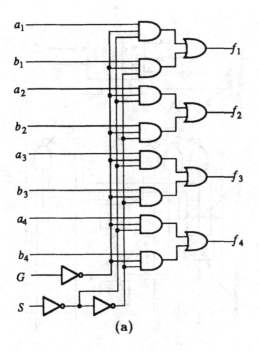

(a)

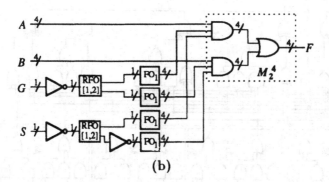

(b)

Figure 2.6: Four-bit 2-to-1 multiplexer: (a) gate-level model M^G; (b) high-level model M^H.

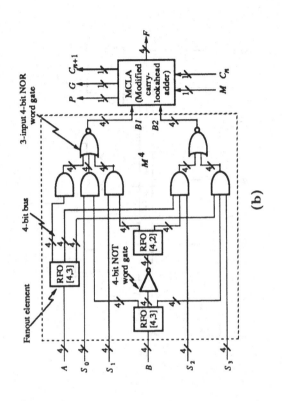

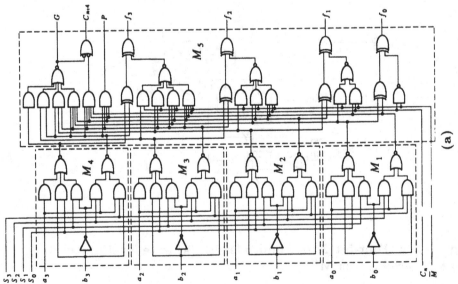

Figure 2.7: The 74181 ALU/function generator: (a) gate-level model M^G; (b) high-level model M^H.

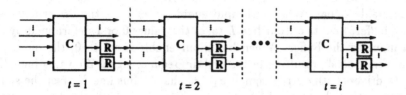

Figure 2.8: Iterative model of sequential circuit at the higher level; C denotes a combinational block and R′ denotes a pseudo-register block.

The generalization of the proposed circuit modeling technique to sequential circuits is quite straightforward, using the standard sequential test generation technique described in Section 1.2.2. The traditional Huffman model for a sequential circuit is assumed, in which registers, and not FFs, form the memory elements at the higher level of representation. As at the gate level, sequential circuit models at the high level normally contain feedback. For test generation purposes, the high-level model containing feedback loops is modified by breaking the loops and constructing an iterative model [Bre76], as shown in Fig. 2.8. As discussed in Section 1.2.2, this iterative model is employed by most traditional gate-level test generation techniques, and thus is in line with our goal of maintaining full compatibility with conventional testing techniques. Furthermore, if the registers follow special design principles like scan design, then the testing problem for sequential circuits can be reduced to that for combinational circuits, and the necessity to construct an explicit iterative model does not arise.

Bidirectional Buses. As mentioned at the beginning of this section, a major problem associated with the construction of high-level circuits models is realistic representation of bidirectional buses. Although standard gate-level circuit models use unidirectional connections only, VLSI circuits often employ bidirectional buses to link the various modules of a system. However, classical test generation techniques require all connections in a circuit to be unidirectional. Bidirectional buses are usually handled by replacing each line in the bus by a pair of unidirectional lines, and modifying the interface logic. Although simple, this solution eliminates important real components such as tristate drivers and receivers (transceivers) from the bus interface. We now develop a unidirectional high-level model of a circuit containing a bidirectional bus which explicitly retains the original interface circuitry, and is compatible with our hierarchical testing methodology.

The sharing of a bidirectional bus like that of Fig. 2.9 depends on the use of tristate drivers whose output signal values are 0, 1 and Z, where Z is the high-impedance state. The state of the output of a tristate bus driver depends on its enable signal. When bus input enable is set to 1, the driver acts as a simple buffer, and its output line L takes the values 0 or 1 determined by the data input to the buffer. When bus input enable is set to 0, the output of the buffer "floats", resulting in the high-impedance signal Z on this line. If n tristate drivers with output signals $x_1 : x_n$ share a bus line L, then the signal x_L representing the state of line L at any time, can be obtained by applying the *connection operation* # [Hay86] to the signals x_i as follows

$$x_L = \#(x_1, \cdots, x_n)$$

For proper operation, at most one driver can place a 0 or a 1 on the bus line L. For example, when all drivers are idle,

$$x_L = \#(Z, Z, \cdots, Z) = Z$$

When driver 1 (with output x_1) is allowed to control the bus, x_L is determined by x_1, since

$$x_L = \#(0, Z, \cdots, Z) = 0$$

or

$$x_L = \#(1, Z, \cdots, Z) = 1$$

The # operation was originally developed for switch-level models, which are inherently bidirectional [Hay87] but, as we show next, can be adapted to higher levels by introducing a new high-level component called the connector element.

A *connector* element is a multiple-input single-output component such that a line in its output bus is obtained by tying together the corresponding lines from its inputs buses. The general case shown in Fig. 2.10, has m input buses and one output bus, such that if vectors $V_1, \cdots, V_m$ are applied to the inputs of the connector element, then the vector $V = \#(V_1, \cdots, V_m)$ appears on the output bus, the operation # being a bit-by-bit connection operation on the vectors $V_1 : V_m$.

Our technique for modeling bidirectional buses is now illustrated by a system consisting of three independent modules interfaced to a bidirectional bus via tristate buffers. For simplicity, we assume that each module has exactly one input bus and one output bus connecting it to the main bidirectional bus, as depicted in Fig. 2.9. Let all bus sizes be n. A high-level model for this system is constructed in three steps as follows. First, the connections to and from the modules are rearranged as shown in Fig. 2.11a, so that all the input buses and all the output buses from the various modules are collected into two separate groups, marked C1 and C2 respectively. Note that in the resulting system, which is functionally equivalent to the original one, signals flow from C1 to C2 only, and not vice versa. The second step consists of modeling

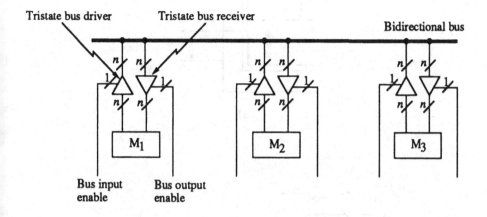

Figure 2.9: Overall organization of a system with bidirectional bus.

C1 by a unidirectional connector element, and C2 by a regular fanout element (RFO), as shown in Fig. 2.11b. Finally, in the third step, our general high-level design technique is applied to each of the modules to obtain their high-level representations.

Observe that a loop similar to a feedback loop in a sequential circuit appears in the final model of the bus system. For test generation purposes, this loop can be handled as in the sequential case, viz., by breaking it along line AA (Fig. 2.11b), and constructing an iterative combinational model from the resulting circuit. Moreover, as in the sequential case, handling of the loop can be considerably simplified by introducing scan-type registers at suitable locations [Ben84].

The simplicity of our high-level models is strongly related to the presence of repeated subcircuits in the gate-level model, and the regularity of their interconnections. With the increasing integration of components, there has also been a trend towards building subcircuits from standard blocks interconnected in a regular manner, because their repetitive structure makes them easier to design [Mea80]. Hence, our high-level modeling technique is particularly useful in generating tests for such fairly regular circuits. In the next section, our modeling technique is applied to a particular class of circuits called k-regular circuits to illustrate its advantages over traditional gate-level methods for test generation purposes.

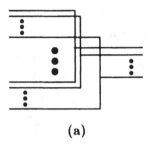

(a)

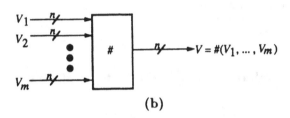

(b)

Figure 2.10: (a) Bus interconnection structure; (b) n-bit m-input connector element.

2.2.2   Fault Model

In order to make the test generation procedure hierarchical in the same sense as the circuit model, the fault models used also need to be hierarchical. The most common fault model for test generation at the gate level is the SSL fault model introduced in Chapter 1. We now define a fault model that is a generalization of the SSL model from single lines to buses. This *bus fault* model, assumes that buses instead of individual lines in high-level circuits can be stuck at constant vector values. The most general kind of bus fault allows arbitrary subsets of lines in a bus to be stuck at 0, stuck at 1, or fault-free. In the VS notation, a fault F on a bus B may be represented by a vector of the form

$$F = b_1 \odot b_2 \odot \cdots \odot b_n$$

where $b_i = 0$ (1) indicates that the line i is stuck-at-0 (stuck-at-1), and $b_i = d$ represents a fault-free line. It is obvious that up to $3^n - 1$ such bus faults can be associated with a bus of size n. However, as will be seen later, only a small subset of the possible bus faults are necessary for test generation. The types of bus faults that have been found to be useful for our purposes are those in which

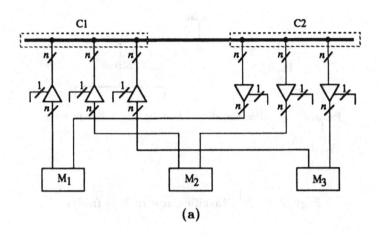

(a)

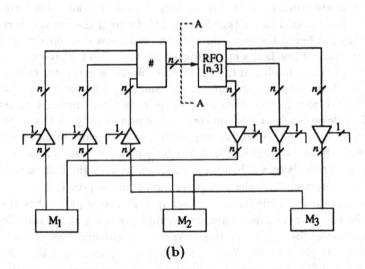

(b)

Figure 2.11: Construction of a high-level model for system with bidirectional bus: (a) rearranging connections to bidirectional bus; (b) introducing fanout and connector components.

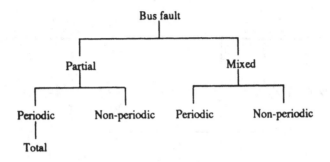

Figure 2.12: Classification of bus faults

the faulty lines in a bus are all stuck at the same logic value. Such faults are termed *partial bus faults*. Of these, the special class of *total bus faults* defined next appear to be most useful. A n-bit bus B is *(totally) stuck-at-0* if all n lines in B are stuck at logic level 0; it is *stuck-at-1* if all n lines are stuck at 1. Obviously, total bus faults reduce to SSL faults if the bus size is reduced to one; thus bus faults are hierarchical in the same sense as our circuit models.

The value of bus faults at the higher level of circuit representation is illustrated by the two models of the 74181 ALU/function generator chip in Fig. 2.7. The gate-level model of the circuit contains 201 lines, and consequently has 402 distinct SSL faults which need to be considered during test generation. The high-level model, on the other hand, contains only 26 buses, and has 52 total bus faults. As we will show in Chapter 3, we can obtain approximately 80 percent fault coverage of SSL faults by generating tests for the 52 total bus faults, which leads to a substantial reduction in both the total number of test patterns generated, and the test generation effort required.

Besides the total bus faults, another special class of partial bus faults called *periodic bus faults* in which lines are periodically stuck at the same logic value are also useful for test generation purposes. A classification of the bus faults of interest is given in Fig. 2.12. The relative importance of the different bus fault types will become evident when we study test generation for these faults in different classes of circuits.

2.3   CASE STUDY: k-REGULAR CIRCUITS

We exploit the presence of repeated subcircuits in the low-level model to construct a high-level model that is easier to test. In this section, we present a detailed application of our circuit modeling approach to the class of k-regular circuits recently introduced by Y. You in [You86,You88]. The best-known examples of such circuits are bit-sliced ALUs and controllers, which are 1-regular, and have long been studied under the heading of cellular or *iterative logic arrays* (ILA's) [Kau67,Fri73]. The concept of k-regularity extends the notion of regularity from such completely regular circuits as ILA's to more general circuits which are not so obviously regular. Intuitively, a k-regular circuit is a one-dimensional array of repeated groups of modules, each group consisting of k distinct module types. Formally defined, a one-dimensional array (cascade) circuit $C_1 : n$ of n modules $C_1, \cdots, C_n$ is said to be k-*regular*if $C_{(i-1)k:ik}$ is isomorphic to $C_{ik:(i+1)k}$, for all $0 < i \leq \lfloor n/k \rfloor$; where $C_{i:r}$ denotes the subcircuit comprising $C_i, C_{i+1}, \cdots, C_r$ and all their interconnections. Some examples of k-regular circuits are provided in Figs. 2.13a–c. Figure 2.13a shows a shifter circuit based on the 74350 IC [Tex85], which is easily seen to be 1-regular. Figure 2.13b shows a parity checker circuit that checks alternately for even and odd parity, and is an example of a 4-regular circuit; this is based on the 74280 IC [Tex85]. Finally, the circuit of Fig. 2.13c checks for even parity only, and is also an example of a 2-regular circuit. Note that the circuits in Figs. 2.13b–c are actually intermediate-level models. As seen from figure, these circuits contain two distinct types of cells, A and B whose internal details are provided in Fig. 2.14. In order to construct $\mathbf{M}^H$ for k-regular circuits, we may sometimes start with such an intermediate-level model instead of a gate-level model because useful structural information is preserved in the former.

The high degree of repetitiveness in the k-regular circuits suggests that the proposed hierarchical test generation technique may be useful in reducing their test generation complexity. In fact, for the special case of 1-regular circuits in which the horizontal and output lines from a module are the same, a high-level model $\mathbf{M}^H$ of the circuit can be easily constructed from the gate-level model $\mathbf{M}^G$, such that a test for a total bus fault in bus B in $\mathbf{M}^H$ detects all SSL faults that can be associated with bus B. The subcircuit consisting of modules $M_{2,1} : M_{2,4}$ in the 4-bit 2-to-1 multiplexer of Fig. 2.6a is an example of such a 1-regular array. Fig. 2.6b shows the corresponding high-level model $\mathbf{M}^H = M_2{}^4$ of this array. The test set presented in Fig. 2.15 detects all total bus faults in $\mathbf{M}^H$. It can be easily verified that it also detects all SSL faults in $\mathbf{M}^G$.

A technique to construct high-level circuit models for general k-regular circuits is presented next, and illustrated with two examples, the ripple-carry adder and the 2-regular parity checker. Simply stated, this procedure forms high-level modules by grouping together identical gate-level modules that are spaced periodically in the array. It forms buses from lines connected to the gate-level modules at intervals of k modules. These buses are now used to connect the high-level modules together, with a loop being introduced to represent the fact that identical modules occurring at intervals of k modules in

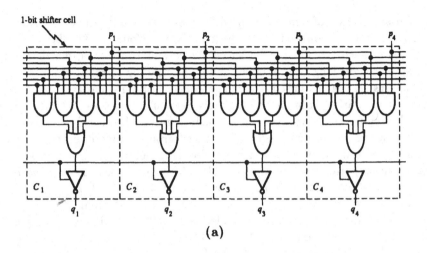

(a)

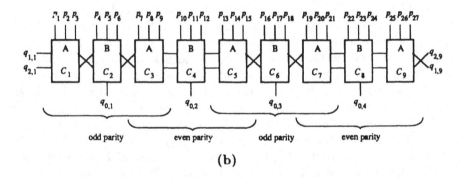

(b)

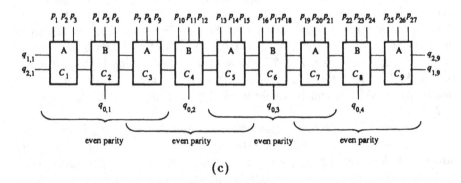

(c)

Figure 2.13: Examples of k-regular circuits: (a) shifter (1-regular); (b) odd and even parity checker (4-regular); (c) even parity checker (2-regular).

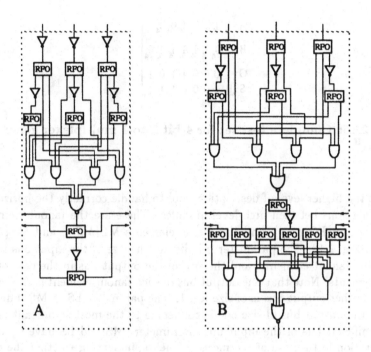

Figure 2.14: Details of cells A and B in Figs. 2.13b–c

the gate-level model are mapped onto the same high-level modules.

Let the general k-regular circuit under consideration consist of n modules, such that $n = qk - r, 0 \leq r \leq k - 1$, where $k \geq 1$ and $n \geq 1$. Following the notation used earlier, $\mathbf{M}^G$ and $\mathbf{M}^H$ denote the low-level and high-level models of the circuit, respectively. The superscripts G and H are also used to differentiate between components or buses of the gate-level and the high-level models. Thus, $\mathbf{M}^G$ for a k-regular circuit consists of modules C_1^G through C_n^G connected to one another as shown in Fig. 2.16. The formal notation used to refer to the various input/output lines connected to the modules C_i^G or C_i^H is also shown in Fig. 2.16. In $\mathbf{M}^G$, the horizontal lines X_{ij} are connected to $X_{(i+1)j}$, and the W_{ij} are connected to $W_{(i-1)j}$. $\mathbf{M}^H$ is obtained by grouping together the identical modules $C_i^G, C_{i+k}^G, C_{i+2k}^G, \cdots$ which are spaced periodically along the array, to form a high-level module C_i^H. The different high-level modules are now connected together using buses of size q or $q - 1$ following the *pseudo-sequential contraction* (PSC) procedure shown in Fig. 2.17. Four special fanout (FO) and merge (ME) components are needed to perform this grouping of lines into

$$\begin{array}{c}A\\B\\G\\S\end{array}\begin{bmatrix}1_4 & 1_4 & 1_4 0_4 & 1_4 0_4 & 1_4 0_4\\0_4 & 0_4 & 1_4 1_4 & 1_4 1_4 & 1_4 0_4\\1 & 0 & 0 0 1 & 1 0 1\\0 & 1 & 0 0 0 & 1 1 1\end{bmatrix}$$

Figure 2.15: Complete test set for 4-bit 2-to-1 multiplexer generated using $\mathbf{M}^H$.

buses at the higher level of description, and to handle correctly the horizontal interconnections between high-level modules. These are the fanout elements FO[MS,n] and FO[LS,n], and the merge elements ME[MS,n] and ME[LS,n] shown in Fig. 2.18. The parameter n indicates the size of the input bus in the case of the fanout elements, and the size of the output bus in the case of the merge elements. Note that one output bus of each fanout element is of size one while the other output bus is of size $n-1$. The parameter LS or MS indicates whether the output bus of size one is connected to the most significant or the least significant line of the input bus. Parameters MS and LS have a similar interpretation in the case of the merge elements, indicating whether the most significant or the least significant line of the output bus is connected to the input bus of size one.

The result of applying the PSC procedure to a 4-bit ripple-carry adder, treated as a 1-regular circuit, is shown in Fig. 2.19. The 1-regularity implies that only one high-level module C_1^H exists in $\mathbf{M}^H$. The number of periods q is obviously four in this case. Moreover, module C_1^G has only one horizontal input and one horizontal output line, the horizontal signal flow being in the same (X) direction throughout the array. Hence, only one ME[$LS,4$] element and one FO[$MS,4$] element are needed for the construction of $\mathbf{M}^H$ from Steps 3 and 5 of the PSC procedure. The interconnections between the high-level module C_1^H and the fanout and merge elements are now easily verified to be as shown in Fig. 2.19b. Note that no ME[$MS,4$] element or FO[$LS,4$] element is required in this case, since the horizontal signal always flows from module C_i to module C_{i+1} in $\mathbf{M}^G$, implying that Steps 4 and 6 are skipped for this example.

An important feature of $\mathbf{M}^H$ for k-regular circuits is the introduction of a loop, although no such loop exists in $\mathbf{M}^G$. $\mathbf{M}^H$ is called *pseudo-sequential* (PS) due to the possible presence of a loops; note, however, that the circuit does not exhibit any sequential behavior in the usual sense. It is easy to see that the pseudo-sequential high-level model contains a constant number of components, whereas the number of components in the gate-level model increases linearly

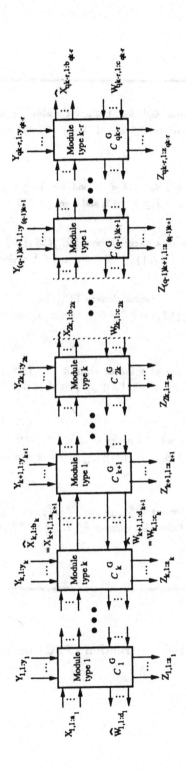

Figure 2.16: Low-level model of k-regular circuit with $n = qk - r$ modules.

1.   Construct high-level modules C_i^H for $1 \leq i \leq c$ where $c = \min(n, k)$. Input/output buses connected to C_i^H are of size q if $i \leq k-r$, or $q-1$ if $i > k-r$.

2.   Connect $\hat{X}_{ij}^H$ to $X_{(i+1)j}^H$ for $1 \leq i \leq k-r-1$, and $k-r+1 \leq i \leq c-1$. Also connect $\hat{W}_{ij}^H$ to $W_{(i-1)j}^H$ for $c \geq i \geq k-r+2$, and $k-r \geq i \geq 2$.

3.   Connect X_{1j}^H to the output of a merge element $ME_{1j}(LS, |X_{1j}^H|)$. Denote the input buses of $ME_{1j}(LS, |X_{1j}^H|)$ I_{1j}^H of size 1, and P_{1j}^H of size $|X_{1j}^H|-1$.

4.   Connect $\hat{W}_{1j}^H$ to the input of a fanout element $FO_{1j}(LS, |\hat{W}_{1j}^H|)$. Denote the output buses of $FO_{1j}(LS, |\hat{W}_{1j}^H|)$ O_{1j}^H of size 1, and $\hat{P}_{1j}^H$ of size $|\hat{W}_{1j}^H|-1$.

5.   Connect $\hat{X}_{(k-r)j}^H$ to the input of a fanout element $FO_{nj}(MS, |\hat{X}_{(k-r)j}^H|)$. Denote the output buses of $FO_{nj}(MS, |\hat{X}_{(k-r)j}^H|)$ O_{nj}^H of size 1, and $\hat{P}_{nj}^H$ of size $|\hat{X}_{(k-r)j}^H|-1$.

6.   Connect $W_{(k-r)j}^H$ to the output of a merge element $ME_{nj}(MS, |W_{(k-r)j}^H|)$. Denote the input buses of $ME_{nj}(MS, |W_{(k-r)j}^H|)$ I_{nj}^H of size 1, and P_{nj}^H of size $|W_{(k-r)j}^H|-1$.

7.   If $r \neq 0$ and $n > k$, then connect $\hat{W}_{(k-r+1)j}^H$ to P_{nj}^H, $\hat{P}_{nj}^H$ to $X_{(k-r+1)j}^H$, $\hat{X}_{cj}^H$ to P_{1j}^H, and W_{cj}^H to $\hat{P}_{1j}^H$. Otherwise, connect $\hat{P}_{nj}^H = \hat{P}_{cj}^H$ to P_{1j}^H, and $\hat{P}_{1j}^H$ to $P_{nj}^H = P_{cj}^H$.

Figure 2.17: Pseudo-sequential contraction (PSC) procedure for k-regular circuits.

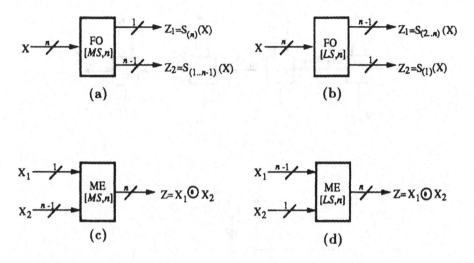

Figure 2.18: The fanout and merge elements used in the PSC procedure: (a) FO[*MS,n*]; (b) FO[*LS,n*]; (c) ME[*MS,n*]; (d) ME[*LS,n*].

with q. The bus size in M^H, on the other hand, increases with q. This trade-off of the bus size for component count implies a significant reduction in the number of components in M^H compared to M^G, even for moderately large values of q, and will be shown in the next chapter to be of great value in reducing the test generation complexity. The only problem of using the pseudo-sequential model directly for test generation is the presence of loops in it, because the standard test generation algorithms require acyclic circuit models. However, we can solve this problem in the standard way discussed in Section 1.2.2 by breaking the feedback loop in the PS model, thereby converting it to an acyclic *modified pseudo-sequential* (MPS) model. The MPS model of the ripple carry-adder is shown in Fig. 2.19c. A pair of input/output buses is created for each loop that is broken, and these buses are referred to as the *pseudo-state input* (PSI) and *pseudo-state output* (PSO) buses.

A major advantage of the MPS (and PS) model over other high-level models discussed in Chapter 1 is that the number of components it contains is constant for any the array size n if $n \geq k$. The usefulness of the MPS model will be demonstrated further in the next chapter where we show that a test for a total bus fault on a bus B in the MPS model, if it exists, detects all SSL faults associated with bus B. However, it will also be shown that it may not be possible to generate tests for all total bus faults in the MPS model, even though the SSL faults on the corresponding lines in the gate-level model are detectable. Hence, sometimes we need to consider another class of bus faults

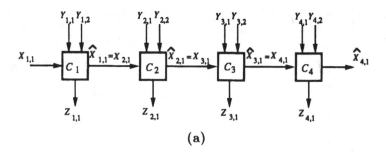

(a)

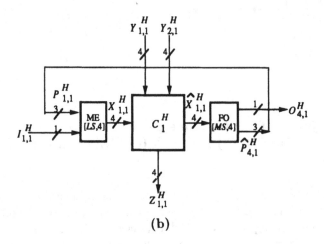

(b)

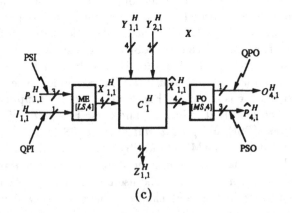

(c)

Figure 2.19: (a) 4-bit ripple-carry adder as a 1-regular array; (b) its high-level pseudo-sequential (PS) model; (c) its modified pseudo-sequential (MPS) model.

called *periodic bus faults* for k-regular circuits, which, when used in conjunction with the MPS models, generate tests with even better SSL fault coverage.

A bus B of size n is said to be *periodically stuck-at-0* or *stuck-at-1* if lines $i, p+i, 2p+i, \cdots$ are stuck at logic level 0 or 1, for some integers p and i, such that $p < n$ and $1 \leq i \leq p$. The parameter p is called the *period* of the fault. Periodic bus faults constitute a subclass of partial bus faults; they include total bus faults, which have period $p = 1$, as a special case. In some useful circuits like the the shifter circuit of Fig. 2.13a, SSL faults that cannot be detected by generating tests for total bus faults (because tests do not exist for all total bus faults), can be detected by generating tests for periodic bus faults with a small period, e.g., $p = 2$.

The proposed MPS models for k-regular circuits are truly hierarchical in that we can easily construct a sequence of high-level models $\mathbf{M}^1 = \mathbf{M}^H, \mathbf{M}^2, \cdots, \mathbf{M}^q = \mathbf{M}^G$ for a given k-regular circuit, such that $\mathbf{M}^G$ and $\mathbf{M}^H$ represent the two extreme cases in this sequence of models. To illustrate this hierarchical aspect of the modeling technique further, we first define the array period of a one-dimensional array of modules [You86]. Let $\mathbf{M}^G$ have the following two properties: (1) There exists a $u > 0$ such that for every C_i in $\mathbf{M}^G$ and for $0 < i \leq n - u$, C_i is identical to C_{i+u}; and for all $t < u$, C_i is not identical to C_{i+t}. (2) There exists a $v > 0$ such that for every I_i in $\mathbf{M}^G$ (I_i represents the connection between C_i and C_{i+1}) and for $0 < i \leq n+1-v$, I_i is identical to I_{i+v}; and for all $t < v$, I_i is not identical to I_{i+t}. Then LCM(u,v) $= k_0$, is called the *array period* of $\mathbf{M}^G$. It is obvious that the least value of k such that $\mathbf{M}^G$ can be said to be k-regular is k_0. However, if a gate-level circuit $\mathbf{M}^G$ is k_0-regular, then it is also $k_0 s$-regular for any integer s, $1 \leq s \leq q$.

A sequence of models of the given circuit at various complexity levels can now be obtained via the PSC procedure by considering the given circuit to be $k_0 s$-regular, and varying the parameter s within the permissible limits. For example, a "high-level" model of the 4-bit ripple-carry adder of Fig. 2.19a treated as a 2-regular circuit ($s = 2$) is shown in Fig. 2.20; compare the 1-regular version of Fig. 2.19b. In this case, the MPS model contains two high-level modules, each representing the grouping together of two modules from $\mathbf{M}^G$. The bus sizes for these modules are reduced from four in Fig. 2.19c to two. In general, the bus sizes are the largest when $s = 1$, i.e., when a k_0-regular array is treated as k_0-regular. The bus sizes in $\mathbf{M}^H$ gradually decrease in a $k_0 s$-regular circuit as s is increased. For $s = \lceil n/k_0 \rceil = q$, the bus size becomes one and $\mathbf{M}^H$ reduces to $\mathbf{M}^G$, since the pseudo-state buses vanish in this limiting case.

The advantages of our high-level modeling technique are further illustrated by comparing $\mathbf{M}^G$ and $\mathbf{M}^H$ for the 27-bit parity checker circuit introduced in Fig. 2.13c. The pseudo-sequential model of the circuit appears in Fig. 2.21a, and the corresponding MPS model is shown in Fig. 2.21b. To compare the number of components and buses in the MPS model to those in the gate-level

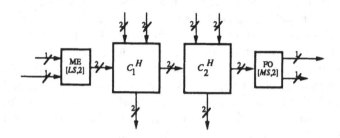

Figure 2.20: 2-regular MPS model of 4-bit ripple-carry adder.

model M^G, we need to expose internal details of the high-level modules A^H and B^H; see Fig. 2.22. M^G contains 228 components which are connected using 359 single-bit lines. M^H for the same circuit has only 51 components interconnected via 91 buses. Thus M^H has only one-fourth the number of components and buses of M^G. Note that four special fanout elements (components 32, 34, 36 and 38) are introduced in M^H which are absent from M^G. The advantages of the proposed approach becomes clearer when M^G consists of q periods, q being an integer. In that case, M^G has $47q$ components, and $81q$ lines. M^H, on the other hand, still contains only 51 components, and 91 buses of size q (or size $q - 1$ for PSI/PSO buses). It will be shown that 90 percent of the SSL faults in the parity checker circuit can be detected with only 15 tests generated for total bus faults in the MPS model using the VPODEM test generation algorithm presented in Chapter 3. In contrast, a test set generated for M^G using the classical PODEM algorithm consists of 60 tests. Moreover, 100 percent SSL fault coverage can be obtained by adding only 16 extra tests to the set of 15 tests mentioned above, resulting in a total of 31 tests for the hierarchical approach, and in a speedup in test generation of approximately three over the conventional approach.

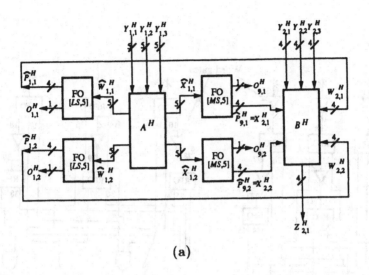

(a)

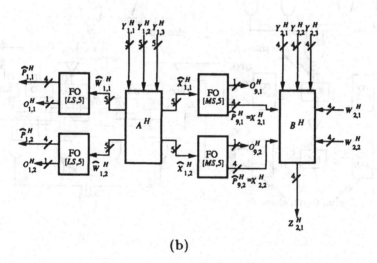

(b)

Figure 2.21: (a) High-level pseudo-sequential (PS) model of parity checker of Fig. 2.13c; (b) corresponding MPS model.

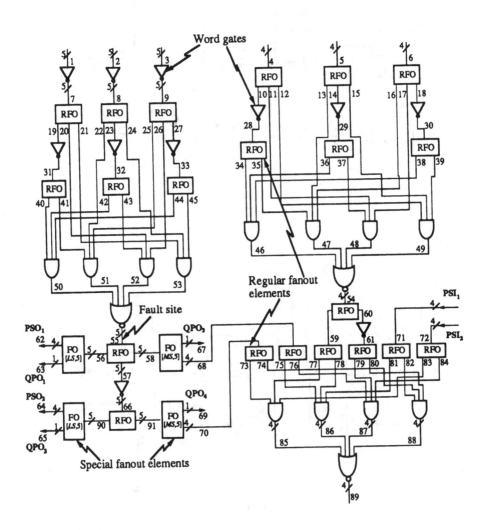

Figure 2.22: Detailed MPS model of parity checker of Fig. 2.13c.

Chapter 3

HIERARCHICAL TEST GENERATION

A test generation procedure is now presented which utilizes the hierarchical circuit and fault models developed in the previous chapter. It aims to exploit the possibility of detecting a significant percentage of the SSL faults by generating tests for total bus faults in the high-level model of a circuit. By effectively merging sets of SSL faults that can be tested in parallel, the number of target faults is reduced, as is the overall test generation effort. Tests for faults that cannot be directly handled can then be obtained by applying the same test generation procedure to a gate-level model of the circuit. Thus, the test generation technique presented here is truly hierarchical, i.e., invariant with respect to the level of the circuit and fault model used, a feature that sets it apart from most other test generation procedures. Moreover, at the gate level total bus faults are the same as SSL faults, implying that the hierarchical test generation technique allows us to obtain complete SSL fault coverage while generating tests for total bus faults only. Experimental results for sample circuits are presented, which show that this approach results in complete test sets for SSL faults that are almost always smaller than those generated by conventional methods.

3.1 VECTOR CUBES

Our hierarchical test generation algorithm allows the high-level circuit model M^H to consist of arbitrary high-level components such as the modified carry-lookahead adder MCLA component used in M^H of the 74181 (Fig. 2.7), as well as simple general-purpose components like word gates. The input/output behavior of general high-level components like the MCLA module are described using the VS notation of Chapter 2, combined with a straightforward generalization of Roth's cubical representation [Rot66], described in

Chapter 1. The elements of the cubes thus obtained are vectors instead of scalars, and hence the cubes are referred to as *vector cubes*. As in the standard (scalar) D-algorithm, vector cubes come in several varieties, which are discussed below.

The *primitive vector cubes* (PVC's) of a component are pairs of VS's that define the component's behavior with respect to vector signals on its input/output buses. For example,

$$1_n \odot X_n \odot X_n / 0_n$$

$$0_n \odot 0_n \odot 0_n / 1_n$$

$$S_{(1..n)}(1_{\lceil n/2 \rceil} \otimes X_{\lceil n/2 \rceil}) \odot S_{(1..n)}(X_{\lceil n/2 \rceil} \otimes 1_{\lceil n/2 \rceil}) \odot X_n / 0_n \qquad (3.1)$$

are three vector cubes of an 3-input n-bit NOR word gate, where X_n represents the don't care vector of size n. Similarly, the following two VS pairs

$$1_4 \odot 1_4 \odot 0 \odot 1 / 1_4 \odot 1$$

$$1_4 \odot 0_4 \odot 0 \odot 0 / 0_4 \odot 0$$

denote vector cubes of the MCLA module in $\mathbf{M}^H$ of the 74181 ALU/function generator.

The concept of cube intersection has also been directly extended from the scalar to the vector case as follows. Two PVC's V_1 and V_2 have a non-empty *vector intersection* V if the corresponding vector elements in V_1 and V_2 are space-compatible, and each primitive element in the first cube can be intersected with its counterpart in the second cube using the rules

$$s \bigcap s = s, \quad s \bigcap X = s \qquad (3.2)$$

where $s \in \{0, 1\}$.

In other words, it is possible to intersect V_1 and V_2 if no pair of corresponding elements in the two vectors, in their fully expanded (atomic) form, belong to the set $\{(1,0), (0,1)\}$. For example,

$$1_n \odot X_n \odot 0_n / 1_n \bigcap X_n \odot 0_n \odot 0_n / 1_n = 1_n \odot 0_n \odot 0_n / 1_n$$

$$1_n \odot 1_n / 0_n \bigcap 0_n \odot X_n / 1_n = \phi$$

where $\bigcap$ now denotes vector intersection.

The *primitive vector D-cubes* or PVDC's of a component are obtained by a modified intersection of the vector cubes of the faulty and fault-free components. In this case, intersection is possible even if a pair of corresponding elements in the two fully expanded vectors belong to the set $S = \{(1,0),(0,1)\}$. These pairs introduce $D/\overline{D}$ values into the resulting PVDC, where D corresponds to $1 \bigcap 0$ and $\overline{D}$ corresponds to $0 \bigcap 1$. Thus, we can obtain the the following PVDC for a 3-input n-bit NOR word gate from the first two vector cubes in (3.1):

$$D_n \odot 0_n \odot 0_n / \overline{D}_n$$

Similarly, the MCLA component has the following PVDC:

$$1_4 \odot D_4 \odot 0 \odot 1/D_4 \odot D \qquad (3.3)$$

Note that in (3.3), an error vector on one input bus leads to error vectors on both output buses. In general, when a high-level component has more than one output bus, an error vector on one input bus may spread to several output buses. In gate-level models, on the other hand, the only way in which an error signal can spread to multiple lines is through fanout of some line, since all gates have only one output. In a high-level circuit model, the fanout element introduced in Chapter 2 is just one of many component types with multiple output buses.

The number of possible vector cubes of a component grows rapidly with increasing bus size. However, it will be seen that we do not have to consider all possible vector cubes if we adopt a hierarchical approach to test generation; it suffices to consider only those cubes relevant to test generation for total bus faults. This follows from the observation that SSL faults that cannot be detected by tests for total bus faults in the high-level model are detected using tests generated for the gate-level model of the circuit, since total bus faults are the same as SSL faults at the gate level. This restriction to total bus faults at all levels of circuit modeling leads to considerable reduction in the number of target faults, as well as a reduction in the overall test generation complexity.

3.2 TEST GENERATION

The question of the SSL fault coverage obtained by generating tests for total bus faults in high-level models of regular circuits is studied first in this section. Problems that arise when dealing with the high-level models of more general circuits are discussed next. Finally, our hierarchical test generation algorithm is presented, and its advantages and disadvantages are discussed.

In order to ascertain the fault coverage of a set of tests, it is first necessary to specify the type of faults to be detected. Although some controversy exists about the suitability of various fault models in approximating physical faults in circuits, the SSL fault model is almost universally used by test generation algorithms. There is considerable experimental evidence indicating that a test set for all SSL faults provides good coverage of the permanent faults encountered in practice [Cas76,Fer88]. Consequently, we also use the SSL model for analysis of fault coverage, i.e., we estimate the fault coverage obtained by the hierarchical test generation procedure in terms of the fraction of detectable

SSL faults in the circuit. The general problem of computing the fault coverage is extremely complex, and we resort to fault simulation techniques to handle it. The problem is much less complicated for regular circuits, i.e., for circuits containing repeated identical subcircuits interconnected in a uniform manner. The fault coverage of some special classes of regular circuits of increasing complexity is therefore studied first.

3.2.1 Repetitive Circuits

The simplest class of regular circuits of interest consist of replicated modules in a one-dimensional array with no intermodule connections. The low-level model M^G of such a circuit composed of n copies of an arbitrary gate-level module C is shown in Fig. 3.1a. To simplify the description while preserving generality, we omit internal details of the module C, and represent it only as a block with single-bit input and output lines, as shown in Fig. 3.1a. Our high-level model M^H for this circuit then consists of word gates and regular fanout elements only interconnected by buses of size n. Again, omitting internal details, this high-level model of the whole circuit is represented simply as a block in Fig. 3.1b, with input/output buses of size n. From [Hay80], we know that if V represents a complete test set for any module C, then $V^{\otimes n}$ is a complete test set for the whole circuit, i.e., for all n copies. However, each column of $V^{\otimes n}$ actually represents a test for a total bus fault in M^H. This leads directly to the following easily proven result:

Theorem 3.1 A complete test set for all total bus faults in a high-level model M^H composed of word gates and regular fanout elements interconnected by buses of fixed size, is also a complete test set for all SSL faults in the corresponding low-level model M^G.

Although practical circuits that conform exactly to the specifications of Theorem 3.1 are rare, many commonly used circuits contain word-oriented subcircuits that satisfy Theorem 3.1. The structure of one such class of circuits is shown in Fig. 3.2a. All the modules $M_{m+1,1}, \cdots, M_{m+1,n}$ are identical, while modules $M_1, \cdots, M_m$ are not necessarily identical to one another. The high-level model of the circuit of Fig. 3.2a appears in Fig. 3.2b. The presence of fanout elements other than regular ones means that the circuit modules are not independent, but no reconvergence of signals from modules $M_{m+1,1}, \cdots, M_{m+1,n}$ is possible with this circuit structure. A simple example is found in the 4-bit 2-to-1 multiplexer shown in Fig. 2.6 It is easy to verify that for such circuits, a complete test set for total bus faults in the high-level model of the circuit is also a complete test set for all SSL faults in the corresponding gate-level model.

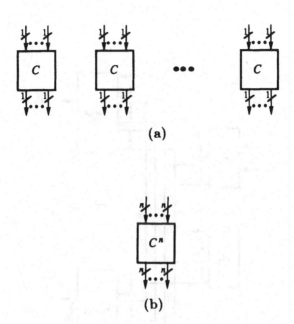

(a)

(b)

Figure 3.1: (a) n identical modules in a one-dimensional array with no intermodule connection; (b) high-level model.

3.2.2 Pseudo-Sequential Circuits

Next, we consider the class of combinational k-regular circuits [You86, You88], which have pseudo-sequential and modified pseudo-sequential models as discussed in Chapter 2. Some important issues in test generation for such circuits are illustrated by a detailed MPS model M^H of a NAND implementation of an n-bit ripple-carry adder, which appears in Fig. 3.3.

The two main issues in generating tests for an MPS model are the fact that the pseudo-state input (PSI) buses are not directly controllable, and the fact that the pseudo-state output (PSO) buses are not observable. The lack of observability implies that simply generating a test for a total bus fault in an MPS model does not guarantee detection of all SSL faults associated with the faulty bus, because the test generation process can lead to the error vector being propagated to an unobservable PSO bus. This situation is illustrated in Fig. 3.3, where the error vector due to the total bus fault F_1 is propagated to PSO_1 (bus 30). However, if line i of bus 10 is s-a-1, $i \leq n-1$, then the generated test pattern merely guarantees the propagation of the resulting error signal to line i of bus 30, which is not an observable output. To ensure the detection of stuck-at-1 faults on all the component lines of bus 10, we now need to verify that the error signal on line i of bus 30 will be propagated to some observable output line when the test pattern is applied to the actual circuit. Drawing

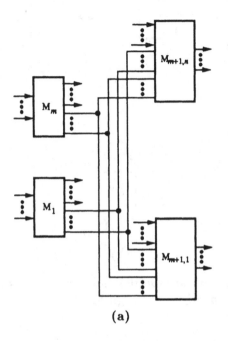

(a)

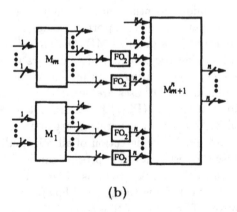

(b)

Figure 3.2: (a) Low-level model containing repeated subcircuits; (b) corresponding high-level model.

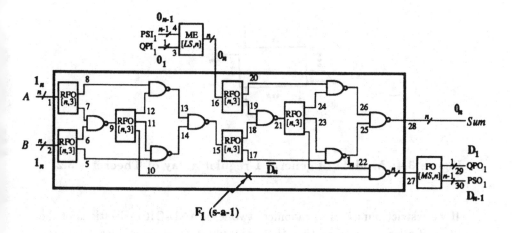

Figure 3.3: MPS model of NAND implementation of an n-bit ripple-carry adder.

on the analogy between the sequential and pseudo-sequential models, we see that this verification may necessitate more than one iteration (pass) through the MPS model to generate a test for a k-regular circuit. These iterations, however, differ significantly from those required for sequential circuits in that all primary input bus assignments found in the first iteration are retained in all subsequent iteration steps. These *pseudo-iterations*, therefore, can only lead to successive refinements of the input assignment found in the first iteration, unlike the case of sequential circuits when iterations lead to a sequence of distinct input patterns.

The second problem of test generation for k-regular circuits is the presence of uncontrollable PSI buses. We must now ensure that any vector assigned to a PSI bus can indeed be applied to the corresponding lines in the actual circuit. An example to the contrary is shown in Fig. 3.3, where the given input pattern applies a vector 0_{n-1} to PSI 13, and generates an output of D_{n-1} at the corresponding PSO 30. This implies an output of 1_{n-1} on PSO 30 in the fault-free circuit. However, since a corresponding pair of lines in the PSI and PSO buses are derived from a single line in the gate-level model, as explained in Chapter 2, they must be assigned the same vector signal in the fault-free case; we designate this the *compatibility requirement*. Two vectors V_1 and V_2, defined on the signal set $\{0, 1, D, \overline{D}, X\}$, are said to be *compatible* with each other, if replacing all occurrences of D and $\overline{D}$ in the two vectors by 1 and 0, respectively, results in vectors V'_1 and V'_2 such that $V'_1 \bigcap V'_2 \neq \phi$. Obviously, vector 0_n is incompatible with D_n, and hence the input assignment shown in Fig. 3.3 must be rejected as invalid.

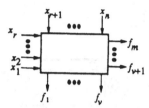

Figure 3.4: Module of general 1-regular array of Theorem. 3.2

If we restrict ourselves to canonical two-level AND-OR realizations of the modules of a 1-regular array, then the compatibility requirement leads to a relatively simple necessary condition to guarantee that tests for total bus faults in the MPS model detect all detectable SSL faults in the gate-level model. Note that the canonical two-level AND-OR (sum-of-minterms) realization of a function is, in general, redundant, i.e., it can contain SSL faults that are undetectable. Hence, the condition derived below is concerned with the detectable SSL faults only. Before presenting this condition, we need to introduce some notation.

Let a module of the 1-regular array have n input lines $x_1, \cdots, x_n$, and m output lines $f_1, \cdots, f_m$, and implement m Boolean functions having the AND-OR canonical form

$$f_j(x_1, \cdots, x_n) = \sum_{i=1}^{p} x_1^{e_{i,j,1}} \cdot \cdots \cdot x_n^{e_{i,j,n}} \tag{3.4}$$

for $1 \le j \le m$, where $e_{i,j,k} \in \{0, 1\}$, and x^1 and x^0 represent x and $\bar{x}$ respectively. Let the first v output lines, i.e., $f_1, \cdots, f_v$, of any module be directly observable (vertical) outputs. A general module employing this notation is shown in Fig. 3.4. Let P be the set of input patterns that can be applied to every module of the 1-regular array in parallel, i.e., the set of input patterns to the MPS model of the circuit that do not violate the compatibility requirement. We call P the set of *parallel tests*. Let $\{V\}$ denote the unordered set of distinct vectors in a VS V. Unordered sets $\{V_j'\}$, which contain tests necessary to detect faults in subcircuits implementing the jth output, are now constructed, for $j = 1, \cdots, v$, following the CHKVCT procedure of Fig. 3.5. Note that $\bigcup(\{V_1\}, \{V_2\})$ represents the normal set-theoretic union of two sets of vectors. $\bigcup_{i=1}^{n}$ is the natural extension of $\bigcup$ to n sets, and $\overline{V}$ represents a VS with the same dimension as V, but whose elements are the complements of those of V.

The construction of the $\{V_j'\}$'s is illustrated in Fig. 3.6 using the example of

```
procedure CHKVCT
   for j := 1 to v do
      Construct V_j = ([e_{1,j,1}]· · · · ·[e_{p,j,1}]) ⊙ · · · ⊙ ([e_{1,j,n}]· · · · ·[e_{p,j,n}])

                    •
      Construct {V'_j} = ⋃{(S_{(1..l-1)}(V_j) ⊙ \overline{S_{(l)}(V_j)} ⊙ S_{(l+1..n)}(V_j))}
                   l=0
   end /*for*/
end /*CHKVCT*/
```

Figure 3.5: Procedure CHKVCT for constructing sets $\{V'_1\}, \cdots, \{V'_v\}$.

a full adder. The canonical Boolean expressions describing the adder's outputs as functions of the inputs are

$$s = a^1 b^0 c_i^0 + a^0 b^1 c_i^0 + a^0 b^0 c_i^1 + a^1 b^1 c_i^1$$

$$c_{i+1} = a^1 b^1 c_i^1 + a^0 b^1 c_i^1 + a^1 b^0 c_i^1 + a^1 b^1 c_i^0$$

Since a full-adder module has only one directly observable output line, i.e., $v = 1$, only one set of vectors $\{V'_1\}$ is constructed by procedure CHKVCT; this set is shown in Fig. 3.6. The set of parallel tests P for a ripple-carry adder is also given in Fig. 3.6. A necessary condition for a test set for total bus faults in the modified pseudo-sequential model to detect all detectable SSL faults in the actual circuit can be now stated.

Theorem 3.2 Let a 1-regular array consist of modules that are canonical two-level AND-OR realizations of switching functions of the form in (3.4). If tests for total bus faults only in the MPS model of the array are to detect all detectable SSL faults, it is necessary that for all j, every vector belonging to $\{V'_j\}$ constructed using the CHKVCT procedure (Fig. 3.5), belong to P.

The proof of the theorem is presented in full in Appendix A; only an outline of the proof is given here. A test T for a total bus fault F in the MPS model of a 1-regular array applies identical inputs to all the modules. Thus, if tests for total bus faults for the MPS model are to detect all detectable SSL faults, it is only necessary to determine if such a test T exists for every total bus fault on an arbitrary bus B in the MPS model whose SSL faults are detectable. To this end, we first identify the set of input patterns that must be applied simultaneously to all modules to cover the detectable SSL faults on the inputs of the AND gates in parallel. Clearly, we know that the VS $\overline{D}_n.1_n$ forms a complete (and minimal) test set for an n-input AND gate. We also know that corresponding to every minterm in (3.4), there is an n-input AND

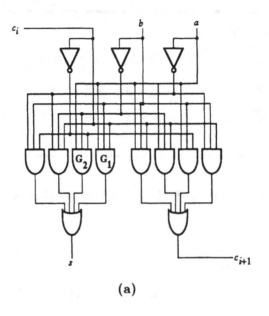

(a)

$$V_1 = \begin{bmatrix} 1 & 1 & 0 & 0 \\ 1 & 0 & 1 & 0 \\ 1 & 0 & 0 & 1 \end{bmatrix}$$

$$\{V_1'\} = \left\{ \begin{bmatrix} 1 & 1 & 0 & 0 \\ 1 & 0 & 1 & 0 \\ 1 & 0 & 0 & 1 \end{bmatrix} \right\} \cup \left\{ \begin{bmatrix} 0 & 0 & 1 & 1 \\ 1 & 0 & 1 & 0 \\ 1 & 0 & 0 & 1 \end{bmatrix} \right\} \cup \left\{ \begin{bmatrix} 1 & 1 & 0 & 0 \\ 0 & 1 & 0 & 1 \\ 1 & 0 & 0 & 1 \end{bmatrix} \right\} \cup \left\{ \begin{bmatrix} 1 & 1 & 0 & 0 \\ 1 & 0 & 1 & 0 \\ 0 & 1 & 1 & 0 \end{bmatrix} \right\}$$

$$= \left\{ \begin{bmatrix} 0 \\ 0 \\ 0 \end{bmatrix}, \begin{bmatrix} 0 \\ 0 \\ 1 \end{bmatrix}, \begin{bmatrix} 0 \\ 1 \\ 0 \end{bmatrix}, \begin{bmatrix} 0 \\ 1 \\ 1 \end{bmatrix}, \begin{bmatrix} 1 \\ 0 \\ 0 \end{bmatrix}, \begin{bmatrix} 1 \\ 0 \\ 1 \end{bmatrix}, \begin{bmatrix} 1 \\ 1 \\ 0 \end{bmatrix}, \begin{bmatrix} 1 \\ 1 \\ 1 \end{bmatrix} \right\}$$

$$P = \left\{ \begin{bmatrix} 0 \\ 0 \\ 0 \end{bmatrix}, \begin{bmatrix} 0 \\ 1 \\ 0 \end{bmatrix}, \begin{bmatrix} 0 \\ 1 \\ 1 \end{bmatrix}, \begin{bmatrix} 1 \\ 0 \\ 0 \end{bmatrix}, \begin{bmatrix} 1 \\ 0 \\ 1 \end{bmatrix}, \begin{bmatrix} 1 \\ 1 \\ 1 \end{bmatrix} \right\}$$

(b)

Figure 3.6: (a) Canonical realization of a full-adder module; (b) sets $\{V_1'\}$ and P (the set of parallel tests) for this adder.

gate in every module of the array. Combining these two pieces of information, it can be shown that the vectors in set $\{V_j'\}$ must be applied to a module to test the AND gates implementing the minterms of f_j in that module. If a vector in $\{V_j'\}$ does not belong to P, then at least one set of AND gates in the modules of the array cannot be tested for SSL faults in parallel (since the necessary inputs cannot be applied simultaneously). Hence, tests for total bus faults cannot detect such SSL faults, even if the individual SSL faults are detectable. Referring back to the example of Fig. 3.6, it is easy to see that vector $\begin{bmatrix} 0 \\ 0 \\ 1 \end{bmatrix} \in \{V_1'\}$, and $\begin{bmatrix} 0 \\ 0 \\ 1 \end{bmatrix} \notin P$, implying that all SSL faults in the canonical two-level AND-OR realization of a ripple-carry adder, implemented using the full-adder module of Fig. 3.6a, cannot be detected by generating tests for total bus faults only in the MPS model of the circuit. However, experimental results will be presented later in this chapter showing that in many commonly used circuits, tests generated for total bus faults in the MPS model detect a large percentage of SSL faults in the circuit, and significantly reduce the total test generation effort.

3.2.3   High-Level Test Generation Algorithm

The problem of obtaining necessary and sufficient conditions for a test set for total bus faults in the high-level model of a circuit to detect all SSL faults in the corresponding gate-level model, is extremely difficult for general circuits. Hence, we have developed a hierarchical test generation algorithm VPODEM which handles both high-level and low-level circuit and fault models of general circuits. It is a substantially extended version of PODEM which recognizes two classes of circuits: combinational, and modified pseudo-sequential; it also treats signals and faults as vectors. While we do not consider sequential circuits explicitly, they could be handled by VPODEM in the conventional manner by constructing iterative array models of the original circuits [Bre76].

The choice of a conventional algorithm as the basis for our hierarchical test generator was motivated by the requirement that the algorithm should reduce to a suitable gate-level test generation algorithm when a gate-level circuit and the SSL fault model are used. As discussed in Sections 2.2 and 2.3, bus sizes become non-uniform at the higher level of representation, and a combinational circuit may be transformed into a pseudo-sequential one. This introduces some new and interesting issues into the test generation process, which require significant changes to the original PODEM algorithm.

The basic high-level test generation algorithm implemented by VPODEM is summarized in Fig. 3.7. It consists of two main procedures, ITERATE and TESTGEN. TESTGEN is a redesigned version of the test pattern generator of conventional (or scalar) PODEM, while ITERATE is an extension to PO-

```
procedure ITERATE
  PSEUDO_ITERATION_COUNT=0
  PSEUDO_ITERATION_FLAG=FALSE
  ENTRY=1
  do while (PSEUDO_ITERATION_COUNT < MAX_COUNT)
    call TESTGEN (ENTRY,PSEUDO_ITERATION_FLAG))
    if (test is not found) then
      if (PSEUDO_ITERATION_COUNT = 0) then
        EXIT in failure
      else
        PSEUDO_ITERATION_COUNT := PSEUDO_ITERATION_COUNT-1
        pop input bus assignment off pseudo-iteration stack
        ENTRY := 2
        if (PSEUDO_ITERATION_COUNT = 0) then
          PSEUDO_ITERATION_FLAG=FALSE
        end /*if*/
      end /*if*/
    else
      if (PSI/PSO buses are absent) then
        EXIT in success
      else
        justify pseudo-state input/output bus assignments
        if (justification is not possible) then
          pop input bus assignment off pseudo-iteration stack
          ENTRY := 2
        else
          push input bus assignment into pseudo-iteration stack
          if (error vector appears on primary output bus) then
            EXIT in success
          else
            PSEUDO_ITERATION_COUNT := PSEUDO_ITERATION_COUNT+1
            assign previous pseudo-state output values to pseudo-state inputs
            assign primary input values from previous iteration
            ENTRY := 3
            PSEUDO_ITERATION_FLAG := TRUE
          end /*if*/
        end /*if*/
      end /*if*/
    end /*if*/
  end /*do while*/
  if (PSEUDO-ITERATION-COUNT = MAX_COUNT) then
    EXIT in success
  end /*if*/
end /*ITERATE*/
```

(a)

Figure 3.7: Pseudo-code of the high-level test generation algorithm VPODEM: (a) ITERATE (b) TESTGEN.

```
procedure TESTGEN (ENTRY,PSEUDO_ITERATION_FLAG)
    do until ((error is propagated to an output bus)
              OR (no alternative input assignment possible))
        call INITOBJ(ENTRY,PSEUDO_ITERATION_FLAG)
/*initobj sets the initial objective if parameter ENTRY is set to 1. If
PSEUDO_ITERATION_FLAG is set to FALSE, then it checks for unini-
tialized condition on the faulty bus. Otherwise, it assumes error signal to
be present on some input, and tries to set an initial objective to propagate
this error signal to some output bus*/
        if ((failure in initobj) OR (ENTRY = 2)) then
/*ENTRY = 2 implies that execution should start with a backtracking
step (alternative assignment to primary input in decision-tree handler)*/
            call DECISION-TREE HANDLER (failure)
            if (no more alternatives exist) then
                EXIT in failure
            end /*if*/
            if (ENTRY = 2) then
                ENTRY := 1
            end /*if*/
        end /*if*/
        if (initial objective is selected successfully by initobj) then
            call BACKTRACE
            call ASSIGN PRIMARY INPUT
            call DECISION-TREE HANDLER (success)
        end /*if*/
        call IMPLICATION
        if (ENTRY = 3) then
            ENTRY := 1
        end /*if*/
/*If ENTRY = 3 to begin with, then initobj does nothing which means all
steps before implication are skipped. Setting ENTRY to 1 signals initobj to
set objective levels in successive cycles*/
        if pseudo-state input and output assignments are not compatible then
            ENTRY := 2
        end /*if*/
    end /*do until*/
    if (error is propagated to an output bus) then
        EXIT in success
    else
        EXIT in failure
    end /*if*/
end /*TESTGEN*/
```

(b)

Figure 3.7: (Contd.)

DEM to handle MPS models. ITERATE and TESTGEN together constitute algorithm VPODEM, which derives its name from the fact that it assigns vectors to buses in a high-level circuit model, in contrast with the conventional PODEM algorithm which assigns scalar values to lines in a gate-level model. The argument ENTRY is used to control the sequence of execution of the three fundamental operations in the algorithm, viz., initial objective selection, implication, and backtracking. The sequence in which these operations are executed in VPODEM may differ significantly from that in PODEM, especially when pseudo-iteration steps are necessary.

The algorithm underlying VPODEM is illustrated by applying it to the parity checker circuit of Fig. 2.13c. A detailed MPS model of the circuit has already been presented in Fig. 2.22. Corresponding PSI's and PSO's in this MPS model are marked by the same subscript. The output lines from fanout elements of type FO[LS,n] or FO[MS,n], which are not PSO's, are designated *quasi-primary outputs* or QPO's to differentiate these buses from primary output buses. In the gate-level model, however, the QPO lines are true primary output lines. Buses 63, 65, 67, and 69 in Fig. 2.22 are examples of QPO's.

The application of VPODEM to the parity checker is illustrated in Fig. 3.8. The fault considered in this example is the total bus fault bus 55 stuck-at-0. Procedure ITERATE begins with a call to TESTGEN with ENTRY set to 1. All bus signals are assumed to be uninitialized, and so have the value X_n assigned to them, where n is the bus size.   The first step in TESTGEN is selection of an "initial objective" [Goe81] consisting of a bus B and a desired (vector) signal V to be placed on B. Thus, initial objective selection is similar to that in conventional PODEM. For the example of Fig. 3.8, the first initial objective consists of bus number 55 (the faulty bus), and the objective level vector 1_5. Backtracing from the fault site to the primary inputs is performed next, leading to the conclusion that assigning the vector 0_5 to bus 1 maximizes the probability of generating an error vector on the faulty bus. This assignment is made and its implications are determined; the bus assignments after the implication step is shown in Fig. 3.8a. (For clarity, most of the bus sizes and numbers in Fig. 2.22 are omitted from this figure.) Primary input buses 2 and 3 are assigned next, using a similar sequence of TESTGEN steps. The resulting vectors assigned to various internal buses are shown in Fig. 3.8b. We see that the first three bus assignments have generated an error vector D_5 on the faulty bus. Compatibility between the various PSI and PSO buses is maintained because buses 71 and 72 are still unassigned, and X_4 is, by definition, compatible with both D_4 and $\overline{D}_4$. TESTGEN terminates in success at this point, since error signals have been propagated to output buses, 62 and 64. ITERATE now calls procedure JUSTIFY to ascertain if the current input assignment has led to a non-X assignment to a PSI while implying X's on the corresponding PSO bus. In this case, no such PSI/PSO pair exists, and JUSTIFY terminates in success. Thus, the first pseudo-iteration step

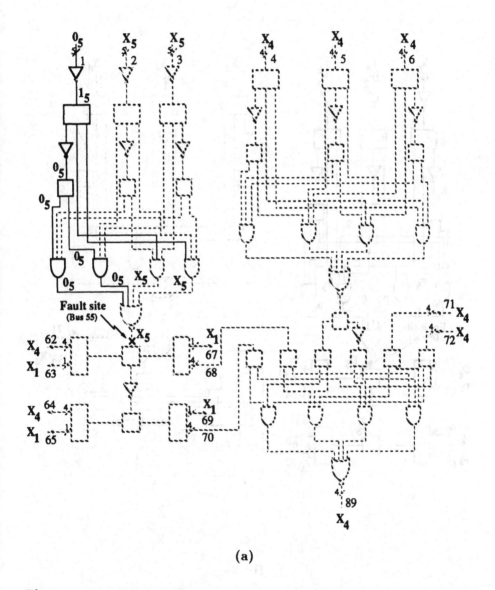

(a)

Figure 3.8: ITERATE and TESTGEN illustrated: (a) after first PI assignment; (b) after assignment of PI's 1, 2, and 3; (c) final assignment to PI's at termination of ITERATE.

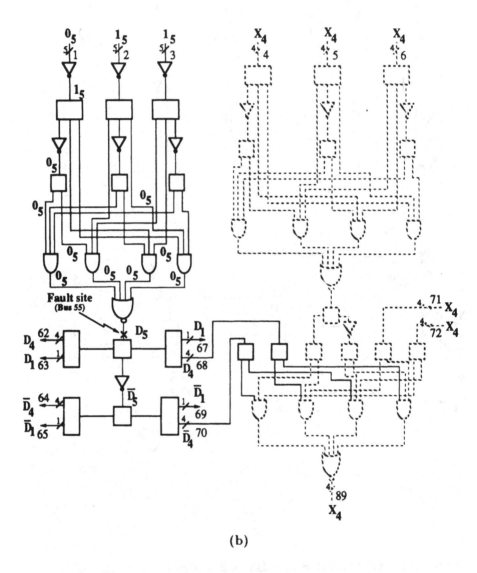

(b)

Figure 3.8: (Contd.)

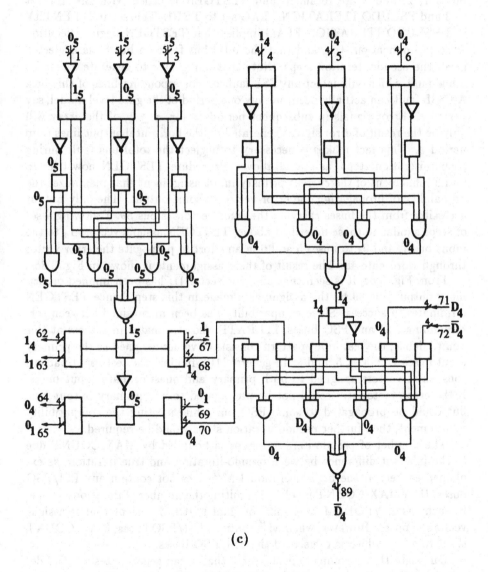

(c)

Figure 3.8: (Contd.)

terminates with the bus assignment shown in Fig. 3.8b.

ITERATE now enters the second pseudo-iteration step by saving the input assignment from the first step, and assigning the vectors D_4 and $\overline{D}_4$ from buses 62 and 64 to buses 71 and 72 respectively. The PI assignments to buses 1, 2, and 3 are retained, and TESTGEN is called with ENTRY set to 3 and PSEUDO_ITERATION_FLAG set to TRUE. This setting of ENTRY and PSEUDO_ITERATION_FLAG implies that TESTGEN begins execution with an implication step, and that the total bus fault on bus 55 is neglected in all the pseudo-iteration steps. Note that our goal is to generate a test for a bus fault which will detect any SSL fault on corresponding lines of this bus. An SSL fault can actually occur in only one period of the gate-level model, say period s. Error signals in subsequent periods $s + r$, $r \geq 1$, of the array will only be the result of error signals generated on the horizontal output lines from period s. This fact makes it necessary to neglect the total bus fault during pseudo-iteration steps, as stated above. Procedure TESTGEN now tries to find a refinement of the current primary input assignment that maximizes the probability of propagating the error vector to some output. The only output accessible from the buses carrying the error vectors is bus 89. Following a set of steps similar to those described above, TESTGEN assigns vectors 1_4 to the input bus 4, and 0_4 to buses 5 and 6 in an effort to propagate the error vector through word gate 48. The result of these assignments is shown in Fig. 3.8c.

From Fig. 3.8c, it is seen that an error vector $\overline{D}_4$ has been implied on primary output bus 89 by the assignments made in this step, hence TESTGEN terminates in success. Since compatibility has been maintained between corresponding PSI and PSO buses, ITERATE also terminates in success at this point. Hence, the test input pattern consists of values assigned to the primary input buses 1, 2, 3, 4, 5, and 6. In general MPS models, the test input pattern consists of vectors assigned to both primary and quasi-primary input buses. If the error vector still could not be propagated to any primary output bus, but could be propagated to some PSO bus while meeting the compatibility requirement, then further pseudo-iteration steps would be required.

The number of pseudo-iteration steps is bounded by MAX_COUNT due to the inherent difference between pseudo-iteration and true iteration, as explained earlier. If the high-level model M^H does not contain any PSI/PSO buses, then MAX_COUNT is set to 1, limiting the number of iterations to one. In such cases, VPODEM works like original PODEM, except that it assigns vectors to buses. However, when M^H contains PSI/PSO buses, MAX_COUNT is set to $q - 1$, which is the size of the PSI/PSO buses.

Our algorithm is complete in the sense that it can generate tests for all detectable SSL faults in any well-formed combinational circuit. Test generation using both high-level and gate-level models may be necessary to obtain complete fault coverage. However, for circuits containing regular subcircuits like k-regular circuits, a large percentage of SSL faults are expected to be detected by tests derived from the high-level model, as suggested by the following result:

Theorem 3.3 A test for a total bus fault $F_e = e_n, e \in \{0, 1\}$, on a bus B in the MPS model of a k-regular circuit generated using VPODEM detects all SSL faults $f_e = e_1$ on individual lines of B in a gate-level model of the circuit.

A complete proof of the theorem is presented in Appendix B; only an informal outline is given here. First, note that VPODEM finds a test only if ITERATE exits in success. The compatibility check in TESTGEN ensures that any test pattern generated for the MPS model can be applied to the original (unmodified) circuit. Given that an MPS circuit model is being considered, ITERATE terminates in success only if an error vector appears on a primary output bus, or PSEUDO_ITERATION_COUNT = MAX_COUNT. The latter case will be dealt with separately, while we concentrate on the case where an error vector is propagated to a primary output bus during some pseudo-iteration step. As mentioned earlier, the horizontal signal flow in the array model—gate-level as well as MPS —is assumed to be unidirectional.

First, we consider the case when the signal flow between modules i and $i + 1$ in is from left to right for all i.

Case A: Let the test pattern generated for the total bus fault $F_e = e_n$ in the MPS model $\mathbf{M}^H$ be T. If an error vector appears on a primary output bus PO in the first iteration step, i.e., with PSEUDO_ITERATION_COUNT = 0, then an SSL fault $f_{je} = e_1$ on line j of bus B in the gate-level model $\mathbf{M}^G$ leads to an error signal on line j of PO, implying that all such faults in $\mathbf{M}^G$ are detected by T. Next, suppose that the error vector is propagated to some pseudo-state output bus PSO_0 in the first iteration step, and is propagated to the primary output bus PO in the next step. In this case, the error signal due to f_{je} in $\mathbf{M}^G$ is propagated to line j of bus PSO_0, and to line $j + 1$ of PO, implying that the SSL fault f_{je} can be detected by T. In general, consider the case when the error vector due to f_{je} is propagated to pseudo-state output buses $\text{PSO}_0, \text{PSO}_1, \cdots, \text{PSO}_m - 1$ for m cycles, $m < \text{MAX\_COUNT} - 1$, and is then propagated to a primary output bus PO in the $(m+1)$th cycle. In this case, the error signal due to f_{je}, $1 \leq j \leq \text{MAX\_COUNT} - m$, is propagated to line $j + m$ of bus PO when T is applied to it, implying that such SSL faults are detected by T. On the other hand, if $\text{MAX\_COUNT} - m + 1 \leq j \leq \text{MAX\_COUNT}$, then the error signal due to f_{je} is propagated to the horizontal output line of the nth module corresponding to bus $\text{QPO}_{\text{MAX\_COUNT}-j}$, when T is applied to $\mathbf{M}^G$. Since the horizontal output of the nth module is observable, the f_{je}'s in $\mathbf{M}^G$ are detected in this case too by T.

If ITERATE terminates when PSEUDO_ITERATION_COUNT = MAX_COUNT, the error signal due to a SSL fault f_{je} in $\mathbf{M}^G$ is propagated to a horizontal output line corresponding to the bus $\text{QPO}_{\text{MAX\_COUNT}-j}$, when T is applied. Thus, for all types of termination, SSL faults f_{je} in $\mathbf{M}^G$ affecting

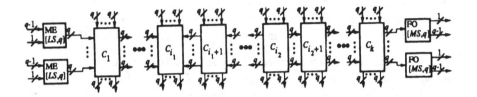

Figure 3.9: MPS model of k-regular circuit in which horizontal signal flow direction varies

lines corresponding to those in bus B of the MPS model $\mathbf{M}^H$ are detected by the test T generated for total bus fault F_e on B.

Case B: Now consider the case where the direction of horizontal signal flow between modules i and $i+1$ varies with i. Since the interconnections are repeated every k modules in $\mathbf{M}^G$, it suffices to consider the direction of signal flow in the first k modules of the array. Note that the directions of signal flow in various buses in $\mathbf{M}^H$ correspond exactly to those in the first k modules of $\mathbf{M}^G$. Let the horizontal signal flow be left to right for modules $1:i_1$, right to left for modules $i_1+1:i_2$, and left to right again for modules $i_2+1:k$, as illustrated in Fig. 3.9. We consider the faults in the three groups of modules separately.

- **Subcase B.1:** First, consider total bus faults in modules $1:i_1$. ITER-ATE can terminate in success for these faults only if the error vector gets propagated to one or more PO buses connected to modules $1:i_1+1$ in the first pseudo-iteration step. The error vector cannot be propagated to any PSO bus in this case.

- **Subcase B.2:** Next, consider the case when the total bus faults lie in modules $i_1+1:i_2$. Again, error vectors cannot be propagated to any PSO bus. Hence, ITERATE can terminate in success only if an error vector is propagated to some PO bus in the first pseudo-iteration step.

- **Subcase B.3:** Finally, suppose the total bus faults lie in modules $i_2+1:k$. In this case, the first pseudo-iteration step of ITERATE can terminate in success if an error vector is propagated to either a PO bus or a PSO bus. If the error vector is propagated to a PSO bus in the first pseudo-iteration step, ITERATE executes a second pseudo-iteration step. In this step, the error vector on a PSO bus connected to module k is assigned to a corresponding PSI bus connected to module 1, and all other PI bus assignments are retained. Thus, the second pseudo-iteration step is equivalent to subcase **B.1**, and can terminate in success only if an error vector is propagated to some PO bus.

In all three subcases mentioned above, ITERATE can terminate in success only if an error vector appears on a PO bus. Moreover, there can be at most two pseudo-iteration steps. Now, following arguments similar to those in Case A, we find that SSL faults in k-regular circuits with unidirectional horizontal signal flow between any two adjacent modules are detected by tests generated for total bus faults in the MPS model.

Although Theorem 3.3 has been stated for the specific case of VPODEM, the theorem is essentially independent of the specific high-level test generation algorithm used. This is due to the fact that the fault coverage obtained using the MPS model is independent of the search technique used, as long as the compatibility and justification criteria discussed earlier are met. In fact, we have already reported such a high-level test generation algorithm based on the classical D-algorithm in [Bha85].

The advantages of the new algorithm over conventional test generation algorithms are twofold. First, VPODEM is invariant with respect to the representation level of the circuit and fault models used. We have achieved this invariance by adopting a hierarchical circuit modeling technique, and identifying an appropriate class of faults for the high-level circuit models, viz., total bus faults. Conventional test generation algorithms are, in most cases, tied to the gate-level circuit fault models. The second, and perhaps more important, advantage stems from the hierarchical nature of the proposed test generation technique. Tests can first be generated for total bus faults in the high-level model of the circuit. Due to the grouping of many lines into a single bus at this level, the total number of target faults is significantly less than in a gate-level model of the circuit, leading to considerable reduction in the overall test generation effort. However, experimental results (to be presented in Section 3.3) show that a significant percentage of SSL faults for many useful circuits can still be detected by the test generated this way. Tests can be generated for SSL faults that are not detected by tests derived from the high-level model, using VPODEM and a gate-level model of the circuit, since total bus faults automatically reduce to SSL faults as the circuit model is expanded to the gate level. Thus, 100 percent SSL fault coverage can be obtained using VPODEM, but with considerably less effort.

The hierarchical test generation approach can be extended in a straightforward fashion to handle circuits containing bidirectional buses interfaced via tristate logic. To do this, a model for these subcircuits that uses unidirectional buses only, is constructed along the lines described in Section 2.2. The set of signal values is extended to accommodate two more logic values, viz., the high-impedance state Z, and the unknown values U. From [Ita86], we know that the PODEM algorithm can be modified without much difficulty to handle these extra signals; clearly, similar modifications can be made to TESTGEN.

3.3 IMPLEMENTATION AND EXPERIMENTAL RESULTS

This section outlines a computer program implementing our high-level test generation algorithm VPODEM, and presents experimental results showing the advantages of the proposed algorithm over conventional techniques. The program is written in the FORTRANVS language, which is IBM's version of Fortran 77, for execution by an IBM 3090 mainframe at the University of Michigan. The program consists of 46 procedures (a main program and 45 Fortran subroutines), including two that implement ITERATE and TESTGEN. One of the procedures implements a simple fault simulator ESTCOV which is used in conjunction with the test generator to reduce redundant computation. The choice of Fortran was made from portability, optimization, and possible vectorization considerations. However, the program suffers from one of the major drawbacks of Fortran, viz., the lack of dynamic data structures, requiring the maximum size of circuits it can handle to be predetermined.

3.3.1 Circuit Description

The input to the test generation program is a text file which provides the circuit description, and various user commands to control program execution and output. The input data is assumed to be organized into three blocks. The first block contains component interconnection information in the form of a net list. Buses that are not outputs of any component are marked as input buses to the circuit. Similarly, buses that are not inputs to any component are marked as output buses of the circuit. The next block of input data specifies the sizes and types buses in the circuit. This information is trivial for gate-level models since all buses are of size one. However, bus sizes vary in the high-level models, and this information must be explicitly provided. Moreover, as shown earlier, input/output buses in the high-level model may belong to the class of pseudo-state buses or quasi-input/output buses. Finally, the user options for the program must be provided in the input file. The user must indicate (i) whether the program should generate tests or should be used as a simulator, and (ii) whether fault coverage analysis is to be performed concurrently with test generation.

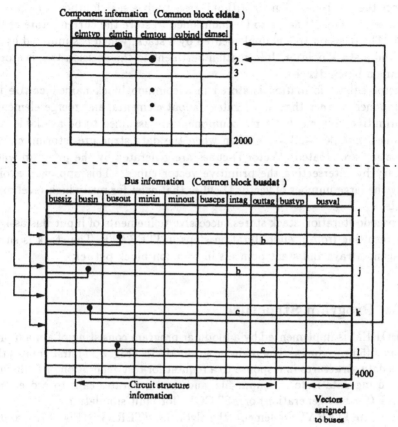

Figure 3.10: Data structure for storing circuit models

3.3.2 Data Structures

The circuit model, either low-level or high-level, is stored in the form of a set of circularly linked lists. This allows bus information for a given component, and component information for a given bus to be extracted with equal ease. Due to the lack of dynamic data structures in Fortran, the lists are simulated using arrays grouped into two common blocks (**eldata** and **busdat**). The organization of the common blocks showing the names of their various fields is presented in Fig. 3.10. The user must provide reference numbers in the input file for each component and bus, which serve as indices to the corresponding lists in the appropriate common blocks. The total number of components and buses, as well as the maximum bus sizes that can be handled are restricted by the sizes of the various arrays used. In our implementation of VPODEM, the maximum number of components that can be handled is 2000, the maximum number of buses is 4000, and the maximum bus size is 16.

Since the test generation algorithm is essentially a search process, a decision or search tree [Goe81] needs to be maintained to store the current state of the search. The decision tree is implemented by a stack, which is simulated by an array and a stack pointer. Information pertinent to the decision tree is stored in common block **dtree**.

Vector cube information is stored in a common block named **ccube** for all components other than word gates, fanout elements, and merge elements. The primitive vector cubes for a component are assumed to be available in a file named 'cubfile', and are read in when the data structures storing circuit information are created. Vector D-cubes are generated by the program when necessary by intersecting the primitive vector cubes. This approach avoids storing the large number of vector D-cubes that exist for many high-level components.

A pseudo-iteration stack stores successive refinements of input bus assignments resulting from pseudo-iteration steps in ITERATE. The stack is simulated using arrays and stack pointers in common block **pstack**.

3.3.3 Program Structure

VPODEM is implemented by a modular program consisting of a main program and 45 procedures. The main program MAIN (Fig. 3.11) first creates the various data structures in response to a request for test generation or fault simulation using data from the input file, and then generates calls to procedures ITERAT (for test generation) or ESTCOV (for fault simulation).

Procedure ITERAT implements the algorithm ITERATE (Fig. 3.7a), and is described by the flowchart in Fig. 3.12. It calls two major procedures TSTGEN and JSTFY. TSTGEN is an implementation of TESTGEN (Fig. 3.7b) which is the modified PODEM procedure that assigns vectors to buses in a high-level circuit model instead of scalars to lines as in a gate-level model. The algorithms underlying procedures ITERAT and TSTGEN have already been described in Section 3.2, and are not elaborated in this section. The structure of TSTGEN is summarized in Fig. 3.13. Procedure JSTFY first finds out if there is any pair of n-bit pseudo-state input/output buses PSI/PSO such that PSI has been assigned a non-X_n value, while PSO is assigned the vector X_n. If it finds such a pair of buses, JSTFY attempts to find assignments to currently uninitialized input buses such that PSO is assigned a non-X_n vector that maintains compatibility with PSI's assignment. This ensures the validity of assignments to uncontrollable pseudo-state input buses. Both TSTGEN and JSTFY call some other important procedures, viz., INITOB, BCKTRC, DTHNDL, and IMPLIC, which are described next.

Procedure INITOB selects an initial objective for the PODEM algorithm. However, since TESTGEN accepts components other than gates, (fanout el-

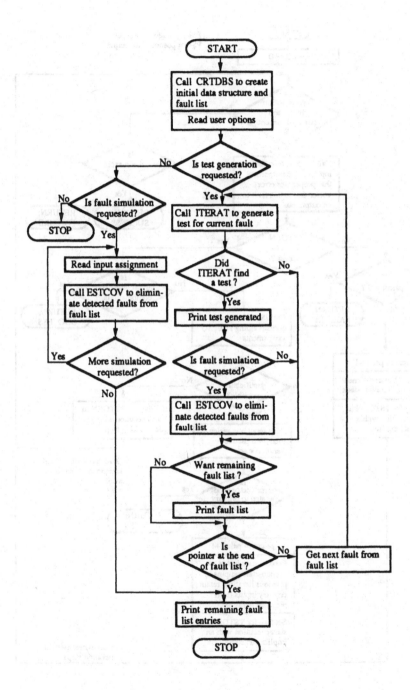

Figure 3.11: Flowchart of main program MAIN.

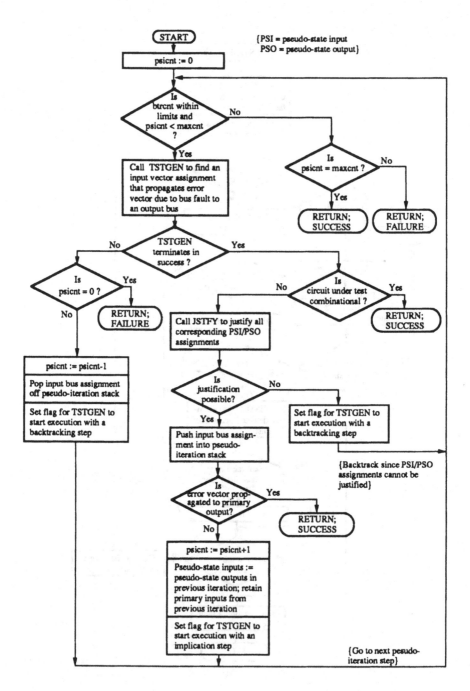

Figure 3.12: Flowchart of iteration control procedure ITERAT.

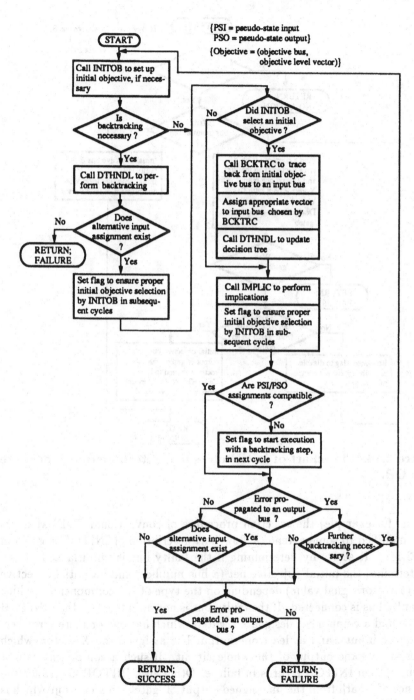

Figure 3.13: Flowchart of vector test generation procedure TST-GEN.

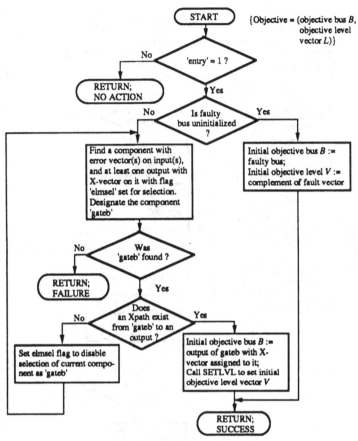

Figure 3.14: Flowchart of initial objective determination procedure INITOB.

ements, for example) the selection procedure of conventional PODEM needs to be modified to some extent. A flowchart description of INITOB is given in Fig. 3.14. It starts by determining if the faulty bus is uninitialized. If so, it determines the initial objective net (a bus number) and an initial objective level (a vector signal value) depending on the type of the component to which the faulty bus is connected. If the faulty bus is not uninitialized, then INITOB tries to find a component designated 'gateb' which has one or more error vectors on its inputs, and at least one output bus assigned the X-vector, which is closest to some output of the whole circuit. If such a component cannot be found, then INITOB returns in failure. Otherwise, INITOB checks for the presence of a path from the unassigned output of 'gateb' to an output which is assigned X-vectors only, by calling procedure XPATH. If such a path is found, i.e., XPATH returns in success, then INITOB sets the initial objective net to be the unassigned output of the component designated 'gateb', and sets the

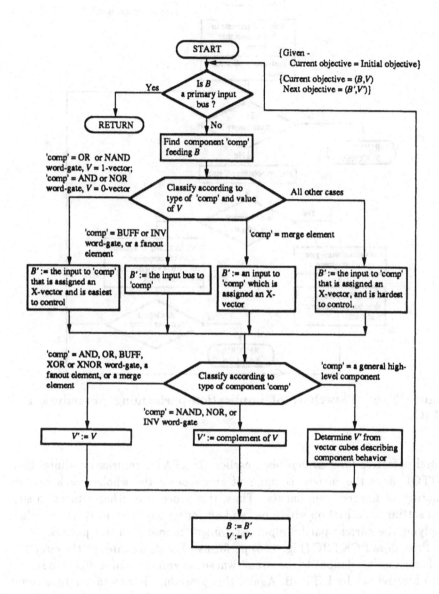

Figure 3.15: Flowchart of backtracing procedure BCKTRC.

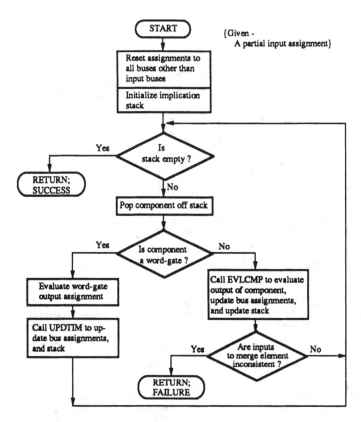

Figure 3.16: Flowchart of implication-performing procedure IMPLIC.

initial objective level as described earlier. If XPATH returns in failure, then INITOB flags the current component and repeats the whole search process ignoring all flagged components. Thus, this procedure either returns in success setting an initial objective net and an initial objective level, or in failure implying the current partial input assignment cannot be a test pattern.

Procedure BCKTRC (Fig. 3.15) performs a backtrace through the circuit to find an unassigned input to the circuit whose assignment is most likely to satisfy the objective set by INITOB. Again, this procedure has to take into account the presence of general components other than word gates in the circuit. It resembles the backtrace process [Goe81] in scalar PODEM to a large extent, except for the modifications necessary to handle components whose behavior is described using vector cubes.

Procedure IMPLIC (Fig. 3.16) performs implications every time a new input bus is assigned, or a previously assigned input bus is assigned a new

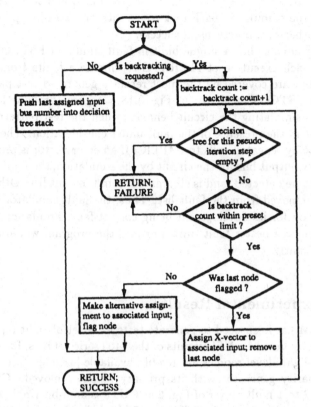

Figure 3.17: Flowchart of decision-tree handler DTHNDL.

vector due to backtracking requested by TESTGEN. The implication process is carried out by first creating a stack of all candidate components whose outputs may need updating. Each component in the stack is then simulated, and the stack of candidate components is also updated during each such simulation cycle. IMPLIC terminates in success if the stack becomes empty, indicating that all implications have been computed. IMPLIC returns in failure only if incompatible assignments are found on the inputs of a merge element $ME[LS,n]$ or $ME[MS,n]$.

Procedure DTHNDL (Fig. 3.17) is the decision-tree handler. It is called either when INITOB or IMPLIC returns in failure, or when a previously unassigned input bus is assigned a non-X_n vector by TESTGEN after successful execution of INITOB and BCKTRC. In the former two cases, DTHNDL performs a backtracking step, and tries out a new partial assignment for the input buses already assigned. In the last case, it adds the new input bus to the decision tree, and stores the vector assigned to it in 'pival'. DTHNDL returns in failure if a backtracking step needs to be performed and the decision tree is empty, or if the cumulative backtrack count 'btrcnt' exceeds a predetermined maximum; otherwise, it returns in success.

The program also has a simple built-in fault simulator ESTCOV, which is called after each execution of ITERATE to eliminate faults from the list of bus faults that are covered by the most recently generated test pattern. The structure of ESTCOV is shown in Fig. 3.18. It goes through the fault list exhaustively, simulating the circuit behavior when the test pattern is applied to the circuit in the presence of the fault under consideration. The simulation is carried out by repeated calls to IMPLIC. If an error vector is propagated to some primary output bus for the circuit by the simulation, the fault in question is assumed to be detected, and is eliminated from the fault list. Otherwise, the fault is not removed from the fault list. Thus, the fault simulator is slow, the major criterion for its development being correct fault simulation, not speed. Incorporation of a better fault simulator into the program would enhance its speed significantly.

3.3.4   Experimental Results

VPODEM has been used to generate tests for a set of eight representative circuits, mostly based on MSI circuits of the 7400 series of IC's, for which high-level and the gate-level models are readily available [Tex85]. Circuit CUT1 is the 74630 parity generator with its primary fanout removed. CUT2 is the 74157 4-bit 2-to-1 multiplexer of Fig. 2.6. CUT3 is an 8-bit ripple-carry adder of the type shown in Fig. 1.9. CUT4 is a 8-bit shifter based on the 74350 IC. CUT5 is the 74181 ALU whose gate-level and high-level models have already been presented in Fig. 2.7. CUT6 is a modified 74381 obtained from [You88]. CUT7 is the parity checker circuit of Fig. 2.22. Finally, CUT8 is a 1/256 decoder tree circuit constructed using modified 1/16 decoders based on the 74154 IC to be described in the next section. The results of applying our test generation program to these eight circuits are tabulated in Fig. 3.19. Figure 3.19a compares the complexity of the gate-level and high-level models using the component count and the number of buses as measures of complexity. Figure 3.19b compares the number of tests generated for total bus faults in the high-level model M^H to the number of tests generated for SSL faults in the gate-level model M^G. It also provides the SSL fault coverage of the tests generated for total bus faults in M^H, and the number of extra tests, if any, needed to obtain 100 percent SSL fault coverage.

From Fig. 3.19a, it is obvious that the high-level modeling leads to a substantial reduction in the circuit complexity. For example, CUT5 has 101 components and 201 lines at the gate level, but only 12 components and 26 lines in the high level model. Similarly, as illustrated in Fig. 2.22, the M^H of CUT7 has

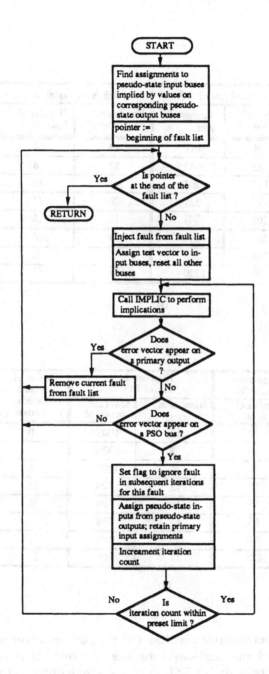

Figure 3.18: Flowchart of fault simulator ESTCOV.

	Gate-level model M^G		High-level model M^H	
	Number of components	Number of lines	Number of components	Number of buses
CUT1	78	144	13	24
CUT2	20	43	12	18
CUT3	120	201	17	30
CUT4	128	227	26	47
CUT5	101	201	12	26
CUT6	240	441	81	164
CUT7	208	359	51	91
CUT8	808	2220	38	47

(a)

	Gate-level model M^G	High-level model M^H		Number of extra tests for 100% SSL fault coverage
	Number of tests	Number of tests	SSL fault coverage	
CUT1	72	12	100%	0
CUT2	16	8	100%	0
CUT3	26	5	100%	0
CUT4	50	12	70%	14
CUT5	54	16	78%	16
CUT6	63	15	77%	21
CUT7	60	15	94%	16
CUT8	483	12	98%	16

(b)

Figure 3.19: Experimental results: (a) number of components and buses in gate-level and high-level models; (b) test set sizes and SSL fault coverage obtained; (c) SSL fault coverage obtained using random test generation.

	Number of tests	SSL fault coverage	
		Tests generated randomly	Tests generated by VPODEM
CUT1	12	88%	100%
CUT2	8	61%	100%
CUT3	5	71%	100%
CUT4	12	69%	70%
CUT5	16	66%	78%
CUT6	15	63%	77%
CUT7	15	81%	94%
CUT8	12	16%	98%

(c)

Figure 3.19: (Contd.)

approximately one-fourth the number of components of its gate-level model. This complexity reduction is perhaps best illustrated by the decoder circuit of CUT8. In this case, M^H contains less than one-twentieth the number of components and less than one-fortieth the number of buses in M^G. Although the size of the buses in the high-level model is usually greater than one, we show next that bus size plays a smaller role in the overall test generation effort than the number of components and buses used.

Figure 3.19b shows that the SSL fault coverage obtained using tests generated for total bus faults in the high-level model is quite high (approximately 80 to 100 percent in most cases). However, the number of tests required to obtain this fault coverage is much smaller than the number of tests obtained using the gate-level model alone. For example, in the case of CUT7, 60 tests were generated using the gate-level model of the circuit, while only 15 tests with 90 percent fault coverage were generated using the high-level model. A set of 16 additional tests were found to cover all SSL faults not covered by the tests generated using the high-level model. Thus, for the circuits considered, the hierarchical approach to test generation leads to only half as many tests as generated by the gate-level technique alone.

VPODEM obviously performs very well for highly regular circuits with small fanout like CUT1, CUT2, and CUT3, or for circuits which have been designed to facilitate high-level test generation as in the case of CUT8. In these cases, test generation at the high level provides complete, or very nearly complete, SSL fault coverage with significantly smaller test sets than are generated using gate-level models only. On the other hand, the presence of complex fanout, as in CUT4 through CUT7, tends to reduce the SSL fault coverage. However, the combined use of both the high-level and the gate-level models is seen to reduce the total number of tests by 50 percent or more, in most cases. The corresponding speedup of test generation, is also significant. For example, the speedup for CUT7 is approximately 3, that for CUT4 is approximately 2, while that for CUT8 is approximately 26.

Finally, we compare our high-level approach to random test generation, which is commonly used as an alternative to algorithmic test generation since pseudo-randomly chosen test patterns are very easy to generate [Wai85]. For each circuit, the number of random patterns generated was made equal to the number of patterns generated using our high-level algorithm, and a standard pseudo-random number generation routine was used to generate the random input patterns. Since the sequence of pseudo-random numbers generated is dependent on the initial seed, three random seeds were used, and the average SSL fault coverage of the three resultant pseudo-random pattern sets was evaluated using the fault simulator ESTCOV in our implementation of VPODEM.

The results of this comparison are shown in Fig. 3.19c. The data shows that the SSL fault coverage of the pseudo-random tests is, in almost all cases, significantly smaller than that obtained using our high-level approach. In fact, only in the case of CUT4, does the random test pattern generation scheme provide fault coverage comparable to that obtained by the high-level scheme, for the same small number of tests. For circuits like multiplexers (CUT2), or decoders (CUT8), the performance of the pseudo-random test generation scheme was found to be far inferior to the proposed high-level scheme. Furthermore, pseudo-random patterns must usually be supplemented with deterministic tests to achieve 100 percent SSL fault coverage with realistic sizes of test sets [Ben84]. In contrast, our hierarchical approach is guaranteed to provide complete SSL fault coverage for arbitrary circuits. The data in Figs. 3.19a–b also show that the test sets from VPODEM are, in general, significantly smaller than those produced by conventional gate-level test generation algorithms.

Chapter 4

DESIGN FOR TESTABILITY

This chapter investigates some design methods aimed at making logic circuits more amenable to testing using the test generation technique proposed in Chapter 3. It is clear from the previous two chapters that this approach is particularly well suited to circuits such as ripple-carry adders and parity checkers which contain repeated subcircuits interconnected in a regular fashion. A bus-oriented high-level model can be easily constructed following the procedure PSC of Fig. 2.17, so that a test set for all total bus faults in the resulting model detects all, or nearly all, SSL faults in the circuit. The scope of our testing methodology can be broadened considerably by formulating some design-for-testability techniques for circuits like carry-lookahead generators, counters, and decoders, etc., which take the form of slightly irregular array or tree circuits. The presence of irregularity makes the construction of high-level models for the unmodified circuits difficult, and complicates test generation. In this chapter, we consider redesigning circuits of this sort to enhance their regularity, and make them better suited to test generation using hierarchical approaches.

We first consider some ad hoc design modification techniques for array-like and tree-like circuits of the kind mentioned above. We then present a systematic design modification technique for a class of tree-like circuits called "generalized" trees [Gaj77,Abr80]. These designs yield fast implementations of functions that have a natural, but slower, iterative implementation. We show that the generalized tree designs can be modified systematically so that tests can be generated from their high-level models with relatively little effort. The chapter ends with a case study of a 16-bit 4-function ALU. A complete set of tests for the modified ALU are derived to illustrate the gains achievable by the proposed design modification schemes.

4.1   AD HOC TECHNIQUES

In this section, we discuss ad hoc DFT techniques for two types of circuits with limited irregularity in their structures: array-like circuits, and tree-like circuits.

4.1.1   Array-Like Circuits

First, we consider design modifications to improve the testability of two general classes of array-like circuits introduced in [You86], viz., inclusively k-regular and nearly k-regular circuits. These classes include such useful circuits as carry-lookahead generators, counters, and encoders. They are characterized by having a somewhat irregular one-dimensional array organization. They are, however, not directly amenable to a bus-oriented high-level model construction in their unmodified form because the array module sizes are not necessarily constant, as seen from the formal definitions presented next. An *inclusively k-regular* circuit consists of n modules $C_1, C_2, \cdots, C_n$ such that the circuit $C_{ik:(i+1)k}$ contains the circuit $C_{(i-1)k:ik}$ as a subcircuit, for $0 \leq i \leq \lfloor n/k \rfloor$, where $C_{j:r}$ represents the circuit consisting of modules $C_j, C_{j+1}, \cdots, C_r$. A *nearly k-regular* circuit, on the other hand, consists of $n+1$ modules $C_1, C_2, \cdots, C_n$ and C_X such that $C_{X,ik:(i+1)k}$ contains $C_{X,(i-1)k:ik}$ as a subcircuit for $0 \leq i \leq \lfloor n/k \rfloor$, where $C_{l,j:r}$ is a circuit consisting of the modules C_l and $C_j, C_{j+1}, \cdots, C_r$. The extra module C_X in a nearly k-regular circuit represents that portion of the circuit that does not occur repeatedly, e.g., control logic. An example of an inclusively k-regular circuit is found in the carry-lookahead generator of Fig. 4.1a, which is inclusively 1-regular. This is easily seen from the fact that a module C_i contains the module C_{i-1} (shaded in Fig. 4.1a for $i = 2$) as a proper subcircuit. An example of a nearly 1-regular circuit derived from a 4-bit carry-lookahead adder is shown in Fig. 4.1b. In this case, module C_X consists of the logic to generate the carry-out signal from the most significant bit. It is also apparent that a module C_i contains the module C_{i-1} (shaded in Fig. 4.1b for $i = 2$), implying that $C_{i,X}$ contains $C_{i-1,X}$ as a subcircuit.

It is worth noting that the k-regular circuits form a proper subclass of the inclusively k-regular circuits, which are, in turn, a proper subclass of the nearly k-regular circuits. Although all these circuits contain some kind of repetitiveness in their structure, the amount of irregularity increases from k-regular circuits to inclusively k-regular ones. The major source of this irregularity is the increasing size of consecutive modules in the array, which makes it difficult to form an MPS model. However, as we show in this chapter, small design modifications can simplify this problem considerably, and allow us to generate tests for such circuits using the hierarchical approach presented in the preceding chapter.

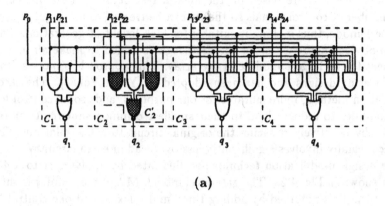

(a)

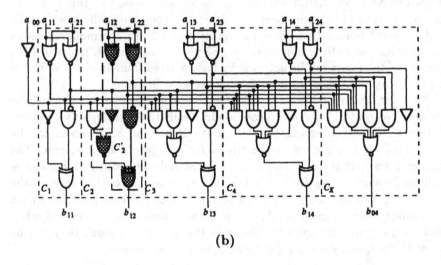

(b)

Figure 4.1: Some array-like circuits: (a) four-bit carry-lookahead generator (inclusively 1-regular); (b) four-bit carry-lookahead adder (nearly 1-regular.)

First, we consider design modifications to inclusively k-regular circuits. These modifications are based on the idea of restructuring a circuit using a small number of control signals so that it can be treated as a k-regular circuit in the test mode, thus making it suitable for high-level test generation. There are two obvious approaches. In the first, logic is added to make all modules equivalent to the largest module in the array. This approach is more suitable for self-testing designs and has been adopted in [You86,You88]. The second modification method, more suitable for our purposes, adds some control logic to all modules to reduce them to their smallest common subcircuit for testing, thereby effectively reducing the original circuit to a k-regular one. This approach usually involves significantly less overhead in extra circuitry.

The design modification technique is illustrated by applying it to a 4-bit counter shown in Fig. 4.2. The gate-level model, M^G of the counter is shown in Fig. 4.2a. It is modified by adding three multiplexers and one control line R, which can be set either to the normal operation mode (R=0) or to the test mode (R=1), as shown in Fig. 4.2b. When R is set to test mode, the circuit is effectively reduced to a 1-regular circuit. Note that the high-level model M^H contains only 15 components and 30 buses, compared to 58 components and 81 lines in M^G. VPODEM was used on an iterative model of this circuit to generate a set of 14 test sequences that detects all total bus faults in an acyclic model derived from M^H, as described in Section 2.2. Fault simulation shows that this test set detects approximately 85 percent of the SSL faults in M^G. The rest of the faults can be detected by tests derived from an iterative model of M^G, using VPODEM if higher fault coverage is desired.

A similar design modification technique can be followed for nearly k-regular circuits. The control logic in this case is added only to the inclusively k-regular portion of the circuit, making it k-regular. Hence, an MPS model can be constructed for this portion of the circuit for test generation purposes. The high-level model of the module C_X can be expected to be identical to its gate-level model since it usually lumps together the irregular control logic in the circuit. C_X can be easily interfaced to the rest of the high-level model via appropriate fanout elements. Again, some SSL faults may be missed when tests are generated for total bus faults in the high-level model; they can be covered subsequently using a gate-level model of the whole circuit.

4.1.2   Tree-Like Circuits

Tree-like circuits such as decoders and decoder trees pose a new problem in constructing high-level models, one that is quite different from that encountered in the array-like circuits. An unmodified $1/2^n$ decoder must be tested for SSL faults by applying all the 2^n possible inputs to it, and appears to have no significant high-level structure. Consider, for example, a typical decoder like the 1/16 decoder found in the 74154 IC; its gate-level model M^G appears

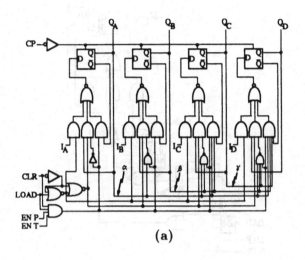

(a)

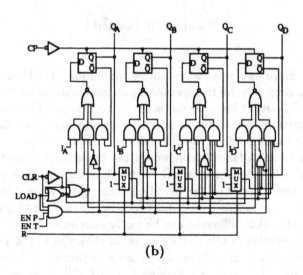

(b)

Figure 4.2: Four-bit counter: (a) gate-level model M^G; (b) modified gate-level model; (c) high-level model M^H of modified circuit in test mode (R = 1).

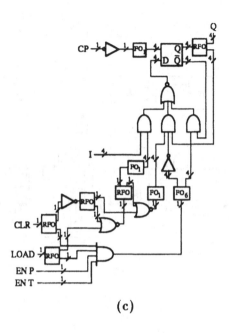

(c)

Figure 4.2: (Contd.)

in Fig. 4.3a. Since any output line can be set to logic level 0 by one specific input combination only, all the possible input combinations must be applied to ensure that the output lines are not stuck at 1. This also implies that no two stuck-at-1 faults on the output lines can be detected simultaneously, suggesting that a high-level model of the decoder must employ output buses of size 1. The complex internal fanout structure of the decoder confirms the suspicion that it contains little useful structure for building high-level models.

A more careful study of the decoder's structure reveals that the various input/output paths in this circuit differ only in the number of inversions they contain, and that this difference can be easily eliminated by the introduction of a set of four exclusive-OR (XOR) gates, as shown in Fig. 4.3b. The extra input R now acts as a control line, which can be either set to 0 (normal mode) or to 1 (test mode). A compact high-level model $\mathbf{M}^H$ of the resulting circuit can be easily constructed when $R = 1$; see Fig. 4.3c. This reduces the number of components from 29 to 19, while increasing the bus size from 1 to 16. The total bus faults in $\mathbf{M}^H$ require 6 tests, which detect all SSL faults in $\mathbf{M}^G$ except faults on inputs to the XOR gates connected to R. An additional 5 tests detect these SSL faults, leading to 11 tests for complete SSL fault coverage, instead of 32 for the unmodified circuit. The test sets for the modified and unmodified designs are shown in Fig. 4.4. The number of circuit components, as well as the number of tests, decreases drastically with larger n. $\mathbf{M}^H$ for the modified

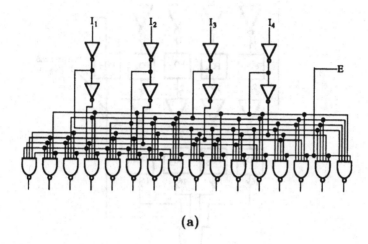

(a)

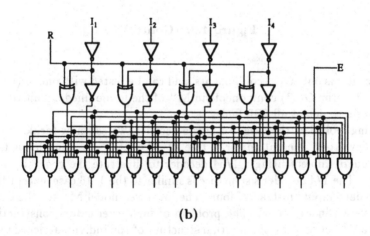

(b)

Figure 4.3: A 1/16 decoder based on the 74154 IC: (a) original circuit; (b) modified circuit; (c) high-level model of modified circuit in test mode (R = 1).

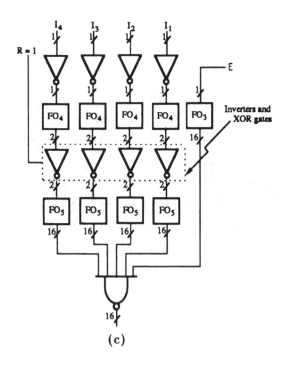

(c)

Figure 4.3: (Contd.)

circuit contains only $O(n)$ components, and can be tested with only $O(n)$ tests, compared to the $O(2^n)$ components in $\mathbf{M}^G$ of the unmodified circuit requiring $O(2^n)$ tests to detect all SSL faults.

Having looked at the case of a single decoder, we can now tackle the issue of high-level test generation for a tree of decoders. We consider a tree of 1/4 decoders implementing a 1/64 decoder to illustrate the general principle involved. The 1/4 decoder's structure is similar to the 1/16 case, except that it has two data inputs instead of four. The low-level model $\mathbf{M}^G$ for the decoder tree is shown in Fig. 4.5a. The problem of high-level model construction is complicated not only by the internal structure of the individual decoders, but also by the fact that the decoders are connected in a tree configuration in which a 1 signal can propagate along at most one branch at any time. One can construct a high-level model of the unmodified tree, as shown in Fig.4.5b, with the indicated changes in bus sizes. However, high-level test generation techniques are mainly useful if a high-level model can be constructed such that, in a large number of cases, error signals can be generated on all lines in a faulty bus in parallel. For this decoder tree, it is easy to verify that the total bus fault B stuck-at-0 cannot be detected in the model $\mathbf{M}_H$ of Fig. 4.5b. In fact, no stuck-at-0 bus fault in $\mathbf{M}^H$ can be detected if the model of Fig. 4.5b

$$
\begin{array}{c}
E \\ I_1 \\ I_2 \\ I_3 \\ I_4
\end{array}
\begin{bmatrix}
1\,1\,1\,1\,1\,1\,1\,1\,1\,1\,1\,1\,1\,1\,1\,1\,1\,0\,0\,0\,0\,0\,0\,0\,0\,0\,0\,0\,0\,0\,0\,0\,0 \\
0\,0\,0\,0\,0\,0\,0\,0\,1\,1\,1\,1\,1\,1\,1\,1\,0\,0\,0\,0\,0\,0\,0\,0\,0\,1\,1\,1\,1\,1\,1\,1\,1 \\
0\,0\,0\,0\,1\,1\,1\,1\,0\,0\,0\,0\,1\,1\,1\,1\,0\,0\,0\,0\,1\,1\,1\,1\,1\,0\,0\,0\,0\,1\,1\,1\,1 \\
0\,0\,1\,1\,0\,0\,1\,1\,0\,0\,1\,1\,0\,0\,1\,1\,0\,0\,1\,1\,0\,0\,1\,1\,0\,0\,1\,1\,0\,0\,1\,1 \\
0\,1\,0\,1\,0\,1\,0\,1\,0\,1\,0\,1\,0\,1\,0\,1\,0\,1\,0\,1\,0\,1\,0\,1\,0\,1\,0\,1\,0\,1\,0\,1
\end{bmatrix}
$$

(a)

$$
\begin{array}{c}
E \\ I_1 \\ I_2 \\ I_3 \\ I_4 \\ R
\end{array}
\begin{bmatrix}
0\,1\,1\,1\,1\,1\,1\,0\,0\,0\,0\,0 \\
1\,0\,1\,1\,1\,1\,1\,1\,1\,0\,0\,0 \\
1\,1\,0\,1\,1\,1\,1\,1\,0\,1\,0\,0 \\
1\,1\,1\,0\,1\,1\,1\,0\,0\,1\,0 \\
1\,1\,1\,1\,0\,1\,1\,0\,0\,0\,1 \\
1\,1\,1\,1\,1\,1\,0\,0\,0\,0\,0
\end{bmatrix}
$$

(b)

Figure 4.4: Test sets for 1/16 decoder of Fig. 4.3: (a) before modification; (b) after modification.

is used, and the only "high-level" model of the unmodified decoder tree that allows the detection of stuck-a-0 total bus faults is M^G itself. This clearly illustrates the problem of applying high-level test generation techniques to an unmodified tree circuit.

The solution in this case is similar to that for the single 1/16 decoder. All the decoders in the tree of Fig. 4.5a are modified by introducing a pair of XOR gates and a control line R. The control lines of the decoders at each level i, are now connected together to form a set of control lines $\{R_i\}$; see Fig. 4.6a. The high-level model M^H for the modified design, shown in Fig. 4.6b, applies in the test mode when all R_i's are set to 1. It is readily verified that 8 tests are needed for all total bus faults in M^H, and that this test set detects all SSL faults in M^G, with the exception of some faults on the input/output lines of the XOR gates. The latter faults can be detected via two tests per level, so that a total of 14 tests are required for the modified design, compared to 68 tests for the original tree.

The above example illustrates the substantial reduction in test generation complexity that can result from appropriate design modifications, even in such apparently complex circuits as decoders and decoder trees.

4.2   LEVEL SEPARATION (LS) METHOD

We now develop a systematic design modification technique (the LS method) for a class of designs which are known to be useful for the imple-

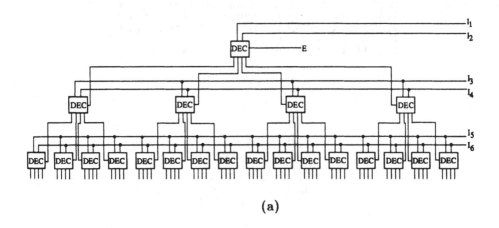

(a)

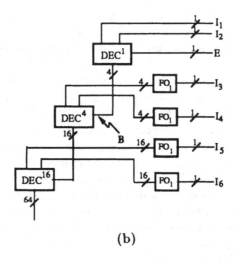

(b)

Figure 4.5: (a) Tree of 1/4 decoders forming a 1/64 decoder; (b) high-level model for the decoder tree.

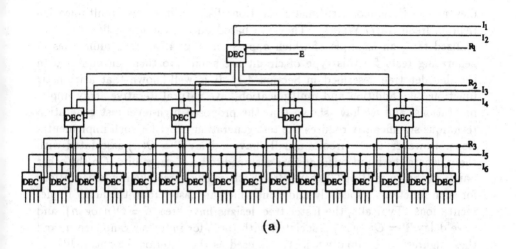

(a)

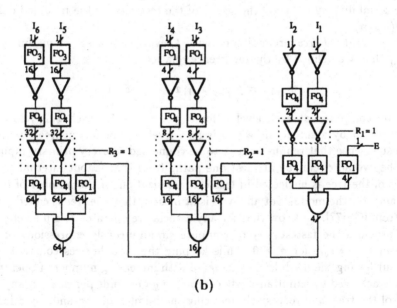

(b)

Figure 4.6: (a) Modified decoder tree; (b) high-level model of modified decoder tree, with control lines R_1, R_2 and R_3 set to 1.

mentation of common arithmetic functions like addition and multiplication [Abr80,Bre80,Gaj77,Wal64]. They are based on a structure called the generalized tree, an example of which appears in Fig. 4.7a. The difficulties in generating tests for this type of circuit are similar to those encountered in the decoder tree described in Section 4.1. It is well-known that arithmetic functions like addition and multiplication have natural iterative array implementations, and we have shown that the proposed high-level test generation technique significantly reduces the test generation effort for such implementations. However, these circuits usually have a delay $\mathbf{T} = O(n)$ associated with them, where the input operands are assumed to be of size n bits. On the other hand, as shown in [Ung77,Abr80,Bre80], there exist tree-like implementations for such functions, especially addition, which are faster than the array implementations. Typically, the faster tree designs have area $A = O(n \log_2 n)$, and have delay $\mathbf{T} = O(\log_2 n)$ associated with them for input operand size n, and these figures are optimal when AT^2 is used as the criterion for optimality. A major problem of the tree designs is that they are much harder to test. In [Abr80], it was shown that, in general, the tree circuits have test sets of size $S = O(n)$, compared to $S = O(1)$ for the iterative designs. In this section, we show that simple modifications to such tree designs can significantly improve their testability by reducing the size S of the test sets for the modified design to $O(\log_2 n)$.

A *generalized tree* circuit is one with l levels, and a set of m_i functions $x_{i+1,j}$ that are defined by the recurrence relation

$$x_{i+1,j} = f_{i+1,j}(x_{i,1}, x_{i,2}, \cdots, x_{i,j}) \qquad (4.1)$$

and are computed at each level i, for $0 \leq i \leq l - 1$. Each $x_{i+1,j}$ in (4.1) is generated by a subcircuit which is an i-level tree or fanout-free circuit in the usual sense that requires $x_{i+1,j}$ to be connected to every primary input of the the overall circuit by at most one path. Note that fanout internal to the nodes of the tree is neglected in this characterization. If the number of input operands for the modules at all levels is at most q, then the tree is called q-ary.

From [Gaj77], we know that for any associative operator $*$, a p-level q-ary tree provides the fastest way to compute simultaneously expressions of the form $e_1 * \cdots * e_{q^p}$, for $p \geq 0$, while keeping the module sizes constant, i.e., without letting the module sizes increase with increasing number of operands. This is achieved by simultaneously combining q operands per node in any one level of the tree, and recursively reducing the number of operands by a factor of q. The process is illustrated for four operands in Fig. 4.8 using a generalized binary ($q = 2$) tree. Here, the modules designated $\mathbf{T}$ perform the operation $*$, while the B modules are buffers whose inputs and outputs have the same logic value. In this case, each node at any level of the tree combines the outputs of at most two modules in the previous level of the tree. Figure 4.7 depicts a useful binary tree implementing 8-bit addition, in which each module output

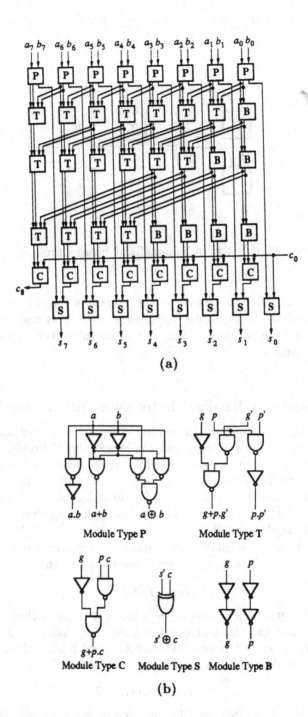

Figure 4.7: Abraham and Gajski's adder design: (a) circuit structure; (b) module definition.

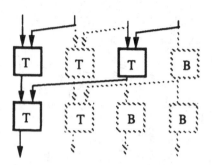

Figure 4.8: A generalized binary tree

$x_{i+1,j}$ corresponds to a pair of lines in the figure. As Fig. 4.7a suggests, tree designs are attractive not only because of their small delay and area, but also because they can be laid out in a regular rectangular geometry suitable for VLSI implementation.

4.2.1   Functions Realizable by One-Dimensional ILA's

The use of generalized trees for fast implementation of functions realizable by one-dimensional ILA's is investigated in [Abr80,Ung77,Bre80]. The principles underlying these designs are derived in [Abr80] based on the observation that a function realizable by a one-dimensional ILA can be described by a first-order recurrence relation. The technique for obtaining the tree design from this recurrence relation, is now reviewed briefly using the formalism developed by Abraham and Gajski [Abr80].

The traditional representation of a first-order recurrence relation consists of a sequence of variables $x_1, x_2, \cdots$ satisfying the equation

$$x_{i+1} = f_i(x_i)$$

where f_i, in general, depends on i. In [Abr80], it is noted that a first-order recurrence relation can also be treated as a set of functions $\mathcal{F} : \mathcal{K} \times \mathcal{X} \to \mathcal{X}$, where $\mathcal{K} = \{k_i\}$ is a set of "constants", and $\mathcal{X} = \{x_i\}$ is a set of variables. The term x_{i+1} is now defined in terms of x_i as

$$x_{i+1} = f_{k_i}(x_i)$$

The members of $\mathcal{F}$ are obtained by choosing different values from $\mathcal{K}$. For example, in the case of a binary adder

$$\mathcal{F} = \{f_{00}, f_{01}, f_{1d}\}$$

$\circ$	f_{00}	f_{01}	f_{11}	f_{10}
f_{00}	f_{00}	f_{00}	f_{00}	f_{00}
f_{01}	f_{00}	f_{01}	f_{1d}	f_{1d}
f_{11}	f_{1d}	f_{1d}	f_{1d}	f_{1d}
f_{10}	f_{1d}	f_{1d}	f_{1d}	f_{1d}

Figure 4.9: Semigroup table for tree design of adder.

where f_{1d} can be treated as f_{10} or f_{11} as necessary. The interpretation of $f_{gp}(c)$ for the adder with $gp \in \{00, 01, 10, 11\}$ is

$$f_{gp}(c) = g + p.c$$

where "+" and "." are the Boolean AND and OR operations, respectively.

Obviously, a one-dimensional ILA constitutes the most natural implementation of a first-order recurrence function, with the horizontal input to module i being x_i, and the corresponding horizontal output being x_{i+1}. However, a faster tree design for the same recurrence relation is obtained if we note that the set of functions associated with a recurrence relation form a semigroup with respect to functional composition operation "$\circ$", which is associative. Moreover, the term $x_i + 1$ in the sequence of terms defined by the recurrence relation can be written as

$$x_{i+1} = f_{k_i}(f_{k_{i-1}}(\cdots(f_{x_0}(x_0))\cdots)) = f_{k_i} \circ f_{k_{i-1}} \circ \cdots \circ f_{k_0}(x_0)$$

where x_0 is a given starting value. As stated earlier, $f_{k_i} \circ f_{k_{i-1}} \circ \cdots \circ f_{k_0}$ can now be implemented by a generalized binary tree circuit, once the composition of any two functions in F is computed. An adder design obtained in this way was already shown in Fig. 4.7. The semigroup table for the elements of F in this case is shown in Fig. 4.9. Independently, using similar techniques, Brent and Kung [Bre80] have presented n-bit adder designs which have area $A = O(n \log_2 n)$ and delay $\mathbf{T} = O(\log_2 n)$.

Although generalized tree designs are faster than array designs, they are usually much harder to test. Previous work [Kau67,Fri73,Sri81a] has shown that one-dimensional ILA's, with at most minor modifications, can be tested by test sets of constant size S, i.e., $S = O(1)$. The generalized trees, on the other hand, require test sets of size $O(n)$ [Abr80]. Moreover, high-level

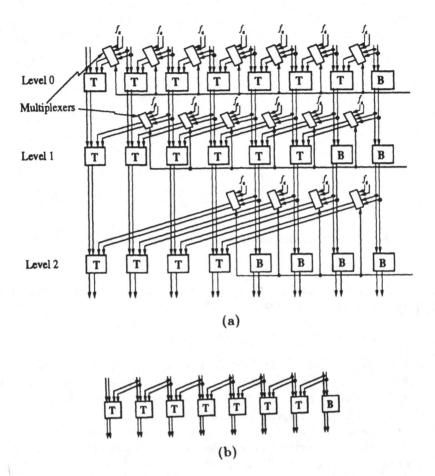

(a)

(b)

Figure 4.10: (a) The LSM method; (b) effect of setting multiplexers in levels 0 and 2 to test mode.

techniques, such as those developed in this book, cannot be used effectively to generate tests for such designs. A systematic modification scheme called the *level separation* (LS) scheme is now presented, which makes generalized trees amenable to testing using our high-level test generation techniques, and also reduces the test set size to $O(\log_2 n)$.

In general, the LS approach introduces a small amount of control logic and some extra input lines to allow a generalized tree to be partitioned into independently controllable and observable levels during testing. This may be done in two ways. In the *level separation by multiplexers* (LSM) approach, multiplexers are inserted into each level i of the tree, and are controlled by a separate control signal R_i; see Fig. 4.10. Each multiplexer has an input f_e which represents the right identity element of the semigroup defined by the

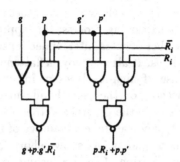

Figure 4.11: The LSR method illustrated for the T modules of Fig. 4.7.

recurrence relation describing the function being implemented. This identity element f_e can always be made to exist by a simple modification of the semigroup in question [Abr80]. Each multiplexer can be set either to the normal mode ($R_i = 0$) when it passes the signal on the main data input to its output, or to the test mode ($R_i = 1$) when it passes f_e to its output. Setting the multiplexers at any level to the test mode makes the corresponding row of modules transparent to data signals. A second implementation of the LS scheme, called *level separation by redesign* (LSR), modifies the **T** module, and introduces control signals R_i and $\bar{R_i}$ (complement of R_i) for every level of the generalized tree, such that when R_i is set to 1 ($\bar{R_i} = 0$) corresponding to the test mode, the **T** modules in the ith level of the tree become transparent to the data signals. Thus, LSR achieves level separation with similar control signals as LSM, but without the introduction of multiplexers. As discussed shortly, the choice between the two implementations of the LS method is determined by the area and time overhead that can be tolerated in a particular application.

Figure 4.10a illustrates the application of LSM to the generalized tree in the adder of Fig. 4.7. In this case, f_e is the input combination 01 to the g' and p' lines of the **T** modules. Thus, with the multiplexers at levels 0 and 2 set to the test mode, the interconnection structure in Fig. 4.10a effectively becomes that of Fig. 4.10b. Modifications to the **T** cell in the adder design when the LSR method is used, are shown in Fig. 4.11.

Multiplexers in the LSM scheme enhance testability by increasing the regularity of circuit structure. This can be contrasted to other DFT techniques that use multiplexers to increase controllability and observability of internal lines in a circuit. In the approach of McCluskey and Bozorgui-Nesbat [McC80] discussed in Section 1.2, for example, the gate-level model of a given circuit is partitioned into subcircuits based on an analysis of functional dependence be-

tween various portions of the circuit. Multiplexers are introduced at the inputs
and outputs of the partitions to allow access to these partitions. In addition to
the complexity of the partitioning process, this approach has the drawback of
introducing a large number of extra interconnections between modules, as well
as some extra control lines. It also employs a flattened gate-level circuit model
which precludes the use of high-level design information to enhance circuit
regularity. The LS method, on the other hand, introduces no extra interconnections between modules in the generalized tree, uses a very small number of
control lines ($O(\log_2 n)$), and can take advantage of high-level design information to enhance circuit regularity. As shown shortly, the restructured circuits
that result from the application of the LS technique to generalized trees have
natural and useful high-level models. A case study, presented in the next
section, will show that high-level test generation algorithms used with these
models can significantly reduce the test set size for generalized tree designs.

Generating tests for a tree modified by the LS method is a three-step process. To simplify its description, we assume $n = 2^k$ for some k. First, tests are
generated for modules in level 0 of the tree, which is easily seen to consist of a
1-regular circuit of **T** cells, and one **B** cell (Fig. 4.12). Referring to Fig. 4.13a,
let $V = v_0 \odot \cdots \odot v_7$ be a VS denoting the tests for the modules in level 0 of the
tree. Obviously, all modules in level 0 can be tested simultaneously by setting
the R_i control signals for all the other levels to the test mode. The second step
in test generation follows from the observation that setting the R_i's to the test
mode in all levels except level i, creates 2^i identical subcircuits, each of which
is also identical to a portion of the circuit in level 0. This is illustrated for $i = 1$
in Fig. 4.13b. Cells belonging to the two different subcircuits are distinguished
by marking them with a † or a ‡. The modules of level 1 can now be tested by
$V_1 = v_3 \odot v_3 \odot v_2 \odot v_2 \odot v_1 \odot v_1 \odot v_0 \odot v_0$. V_1 can be also described in terms
of the VS operation select as $\left(S_{(3..0)}(V)\right)^{\otimes 2}$, where the elements $v_0, v_1, \cdots, v_7$
are treated as primitive elements. In general, the ith level of the tree can be
tested using the VS

$$\left(S_{((n-2^i/2^i)..0)}(V)\right)^{\otimes 2^i} \qquad (4.2)$$

Hence, the second step of the test generation process consists of repeatedly
copying portions of the VS V representing the test set for modules in level 0 of
the tree, as described formally in (4.2). Obviously, no explicit test generation
is required in this step, and the computation time becomes negligible with the
use of proper data structures to store the input pattern sets $v_0, \cdots, v_{2^n}$. In the
third and final step, tests are generated for faults on the control lines (and the
multiplexers if the LSM scheme is used). Again, this can be done one level at a
time, and tests for faults in the control circuitry at the ith level can be derived
from tests for control circuitry at the first level. Thus, for input operands of
size n bits, it is obvious that the test set size S is proportional to the number
of levels in the tree, i.e., $S = O(\log_2 n)$. It is interesting to note that since tests
are generated in a level-by-level manner, the test set size S remains $O(\log_2 n)$,

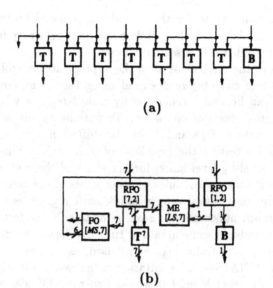

Figure 4.12: (a) Cells at level 0; (b) their pseudo-sequential model.

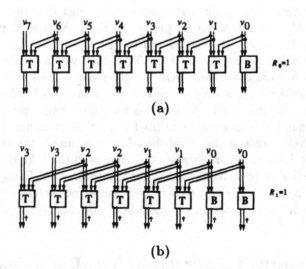

Figure 4.13: (a) Testing level 0 modules of Fig. 4.10 with sets of inputs $v_7, v_6, \cdots, v_0$; (b) testing level 1 modules with sets of inputs v_3, v_2, v_1, v_0.

even if the SSL fault model for the modules is replaced by the more general functional fault model in which a module can have arbitrary functional faults which do not convert it into a sequential circuit.

Application of the tests along with the appropriate control settings is also easy when the tree circuits are designed using the LS scheme. The control signal settings can be easily generated by a shift-register whose input is set to 0 for the normal mode of operation. To put the circuit in the test mode, an appropriate pattern of 1's and 0's can be shifted in to provide the required control settings for testing the first level of T modules. Since the tests are generated on a level-by-level basis, further shifts of this pattern will provide control settings for testing modules in other levels of the tree. This ability to generate the control settings using a simple shift register is especially suited for design environments employing scan paths to enhance testability.

LS design introduces some area and time overhead, the amount of which depends on the specific technology being used, as well as on the complexity of the modules. A NAND implementation of the easily testable 4-bit adder in NMOS technology using Mead-Conway design rules [Mea80] shows less than 15 percent area overhead, and between 5 percent and 19 percent increase in delay, when the LSM scheme is used. Layouts of the T module before and after modification by LSM are shown in Fig. 4.14. The area overhead is around 12 percent, while the increase in delay is expected to be slightly less if the LSR scheme is used. In this case, an approximate estimate of the area overhead is obtained by studying the change in the T module. From Figs. 4.7 and 4.11, it is easy to see that the modified T module contains one more gate than the original 5-gate module. This, combined with the fact that the T modules of the 4-bit adder occupy about 50 percent of the total design, puts the area overhead at 10 percent. For larger adders, the overhead calculated this way, would be closer to 20 percent since the T modules would occupy a significant portion of the total design area. On the other hand, if we take into account details of the layout, it is easy to see that the actual overhead is closer to 15 percent, even for the larger adders. This happens because some unused area in the original T modules can be utilized to partially accommodate the extra gate. If a NOR implementation of the adder is assumed, the area and delay increases for LSM are approximately the same as in the NAND case. A NOR implementation of LSR, on the other hand, introduces a slightly higher area overhead (approximately 20 percent), but almost no increase in delay, since the number of levels of logic in the modules is not changed.

4.2.2   Functions Realizable by Two-Dimensional ILA's

Having illustrated the LS method for the one-dimensional case, we turn to functions such as multiplication that are realizable by two-dimensional ILA's.

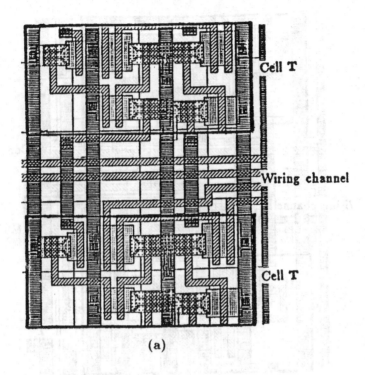

Figure 4.14: Layout of T module: (a) unmodified; (b) after introduction of LSM multiplexers.

This case is inherently more difficult because of the interaction between the signals flowing in the two dimensions. We start by modeling the two-dimensional ILA as an one-dimensional array of "supermodules" in which each supermodule is itself an one-dimensional ILA; see Fig. 4.15. However, the technique for obtaining a generalized tree design that was followed in the one-dimensional case cannot be applied directly to the ILA of supermodules. The input and output signals of the supermodules are vectors whose length is the size of one of the dimensions of the original two-dimensional ILA. For example, the X_i and the K_i inputs to the supermodules in Fig. 4.15b are vectors of size n, implying that the number of distinct combinations that can be applied to the K_i inputs increases exponentially with n. Since the possible inputs to the K_i's now constitute the "constants" [Abr80] in the definition of the recurrence relation, the number of elements in the associated semigroup also increases exponentially with n. Obviously, it is impractical to compute the semigroup operation in this general case, even for moderately large values of n. As a result, practical tree designs can be found only in the case of some special classes of functions realizable by two-dimensional ILA's.

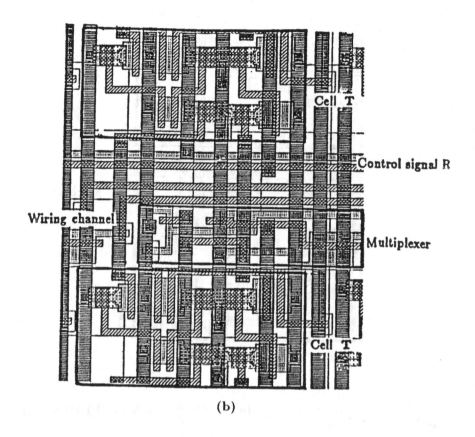

(b)

Figure 4.14: (Contd.)

The class of functions of concern here are characterized by a supermodule output vector X_{i+1} that can be described by a recurrence relation of the form

$$X_{i+1} = f_{K_i}(X_i) = f_{(A_i,B_i)}(X_i) = f(A_i, g(B_i, X_i)) \qquad (4.3)$$

where A_i and B_i are constant vectors. The functions f and g must also have the following properties:

- **P1:** f and g are associative, i.e., $f(X, f(Y, Z)) = f(f(X, Y), Z)$ and $g(X, g(Y, Z)) = g(g(X, Y), Z)$.
- **P2:** g distributes over f, i.e., $g(X, f(Y, Z)) = f(g(X, Y), g(X, Z))$.
- **P3:** $|f| = |g|$, where $|f|$ denotes the size of the output vector produced by f.

Since the supermodules are actually one-dimensional ILA's, it is also reasonable to assume that $|X_{i+1}| = |X_i|$, i.e., $|X_i| = |f|$, and that $|X_i| = |A_i| = |B_i|$. The following result is now seen to hold:

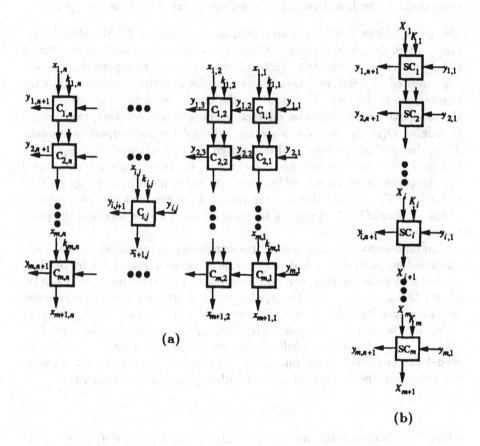

(a)

(b)

Figure 4.15: (a) A two-dimensional unidirectional ILA; (b) corresponding one-dimensional ILA of supermodules.

Theorem 4.1 Let Ω be the class of functions realizable by $m \times n$ ILA's such that the output vector of a supermodule can be described by a recurrence relation of the form (4.3). Then there exists a generalized tree realization of each function in Ω with delay $T = O(\log_2 m \log_2 n)$ and area $A = O(mn \log_2 m \log_2 n)$. Furthermore, this generalized tree design can be modified with $O(\log_2 m \log_2 n)$ extra control signals so that it has a test set of size $S = O(\log_2 m \log_2 n)$.

The proof follows from a two-step realization procedure for the tree design. First, a tree design is constructed from the ILA of supermodules containing $O(\log_2 m)$ levels of "supertree" (**ST**) modules, where each supertree module is implemented by iterative arrays of size $O(n)$. A supertree module is thus the counterpart of the tree (**T**) module in the one-dimensional case, except that the supertree module operates on vector inputs of size n. Next, tree designs containing $O(\log_2 n)$ levels of logic are obtained for each supertree module. Thus, the total delay of the final tree design is $O(\log_2 m \log_2 n)$. $O(\log_2 n)$ control lines are needed for each level of supertree modules, implying that a total of $O(\log_2 m \log_2 n)$ extra control lines are required to modify the design for easy testability. The test set size and the circuit delay are obviously proportional to the number of levels of logic in the final design. Further details of the proof are given in the appendix.

For functions like multiplication, the semigroup operation can be effectively computed in constant time and in space that grows as n instead of $n \log_2 n$. This implies that such functions can be implemented by a tree circuit with a delay of only $O(\log_2 m)$ instead of $O(\log_2 n \log_2 m)$, and area of $O(nm \log_2 m)$ instead of $O(nm \log_2 n \log_2 m)$. To see this, consider a $n \times n$ carry-save multiplier whose inputs are two unsigned binary numbers $A = (a_{n-1}, \cdots, a_0)$ and $B = (b_{n-1}, \cdots, b_0)$. Such a multiplier is shown in Fig. 4.16a for $n = 4$. If we model the two-dimensional multiplier array as an one-dimensional array of supermodules, the function of the $(i + 1)$th supermodule is found to be

$$f_{K_i}(X_i) = X_i + K_i$$

where "+" denotes n-bit arithmetic addition, and K_i is a shifted version of $A.b_i$. The semigroup recurrence operation o for the ILA of supermodules is

$$f_{K_i} \circ f_{K_{i-1}} = f_{K_i + K_{i-1}}$$

Thus, we can implement the semigroup operation for multiplication in constant time and in space that grows only as n by using carry-save adders. The final multiplier design, with the control signals introduced to make it easily testable, is shown in Fig. 4.16b for $n = 4$. The test generation procedure for the tree design of the multiplier is similar to that of the one-dimensional case; the resulting test set size for the tree design of the multiplier grows only as $\log_2 n$.

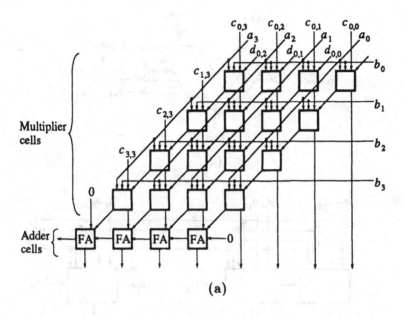

(a)

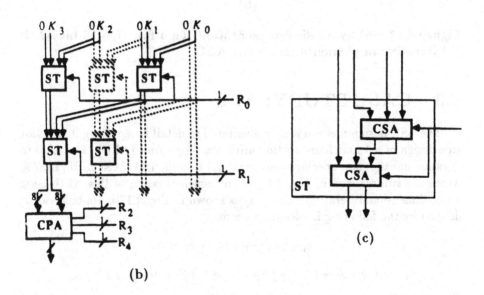

(b)

(c)

Figure 4.16: (a) A 4×4 carry-save multiplier; (b) tree implementation of the multiplier; (c) cell type ST.

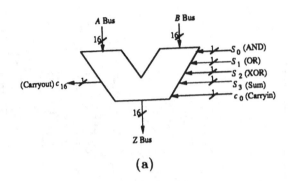

(a)

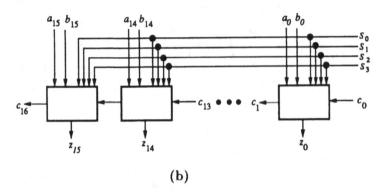

(b)

Figure 4.17: (a) Symbolic representation of a 4-function 16-bit ALU; (b) iterative implementation of the ALU.

4.3   CASE STUDY: ALU

The LSR design method is now illustrated in detail by applying it to a fast tree design of a 16-bit four-function arithmetic-logic unit (ALU). The ALU in question has the same specifications as the ALU in the IBM 3125 CPU [Tan76], which are summarized in Fig. 4.17a. An implementation of this ALU using a one-dimensional iterative logic array is shown in Fig. 4.17b. Its behavior is defined by the following Boolean equations

$$c_{i+1} = (a_i + b_i).c_i + a_i.b_i$$

$$z_i = (a_i.b_i).S_0 + (a_i + b_i).S_1 + (a_i \oplus b_i).S_2 + (a_i \oplus b_i \oplus c_i).S_3$$

Obviously, the recurrence relation describing c_{i+1} in terms of c_i is exactly the same as that of a binary adder, and hence the same semigroup table of Fig. 4.9 applies.

A generalized tree design for the ALU can now be obtained in a straightforward manner following the procedure used for the adder of Fig. 4.7. It employs

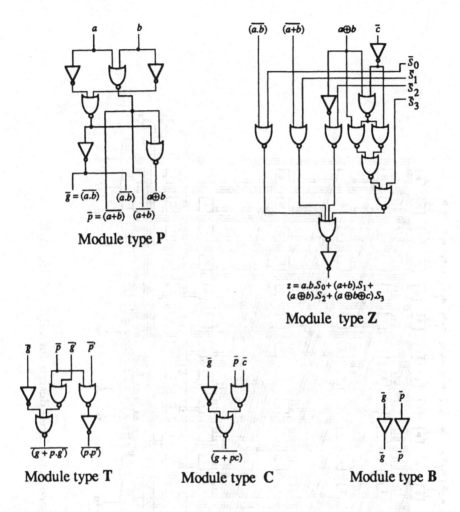

Figure 4.18: Module types in the tree design for ALU of Fig. 4.19.

four types of modules designated **T**, **C**, **P**, and **Z**. The tree (**T**) modules and the carry-generating (**C**) modules are also found in the binary adder. The preprocessing (**P**) modules and the postprocessing (**Z**) modules generating the z_i's, however, are different from the adder case. NOR realizations of various modules appear in Fig. 4.18. The overall organization for the unmodified tree design of the ALU is shown in Fig. 4.19.

As in the case of the adder, the interconnections between the **T** modules pose the major problem in applying our high-level test generation technique to the ALU. The LSR scheme chosen to make this ALU easily testable, necessitates modification of the **T** module logic, and the introduction of a control

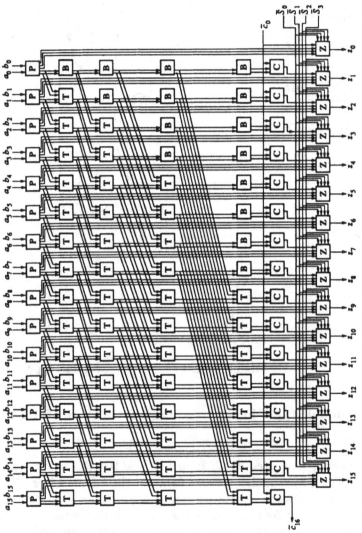

Figure 4.19: Tree design for ALU of Fig. 4.17.

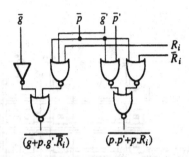

Figure 4.20: Modified T module

line R_i for each level in the tree. The choice of LSR over LSM is motivated by the fact noted in Section 4.2 that for a NOR implementation of the modules, the degradation in speed due to the modification is negligible. The resulting modified **T** module is shown in Fig. 4.20, and the final modified tree design appears in Fig. 4.21. It is estimated that for a NOR implementation of the modules in NMOS technology using Mead-Conway design rules, the area overhead resulting from the design modifications is less than 20 percent.

The advantage of the proposed design scheme is now seen by comparing the test derivation process and the test set sizes of the original and the modified designs. Using the VPODEM test generation algorithm, and a detailed MPS model of the circuit consisting of the **P** cells, the first level of **T** cells, the **C** cells, and the **Z** cells [Bha88], we find that this circuit can be tested for all detectable total bus faults with a test set of size 40. Fifteen of these tests are required to detect faults in the **T** modules, while the rest are required for faults in the **P**,**C**, and **Z** modules. An additional 48 tests must be generated to detect SSL faults in the **T** modules of the first level that are not detected by the tests for the total bus faults. From (4.2), we see that the whole tree, i.e., the portion of the design consisting of the **T** and **B** modules, can be tested for all SSL faults with 277 tests, of which only 88 need to be generated explicitly. The test set size for detecting all SSL faults in a gate-level model of the unmodified tree design, on the other hand, is greater than 400.

The fact that tests need only be generated for the first level of **T** modules also results in a significant reduction in the overall test generation effort due to the considerable reduction in the number of components in the circuit model for which tests are generated explicitly. The unmodified tree design for the ALU contains more than 900 components and almost 1400 lines. After modification, however, the high-level model needed for test generation contains only 90 components and 138 buses. Furthermore, approximately 90 percent of the

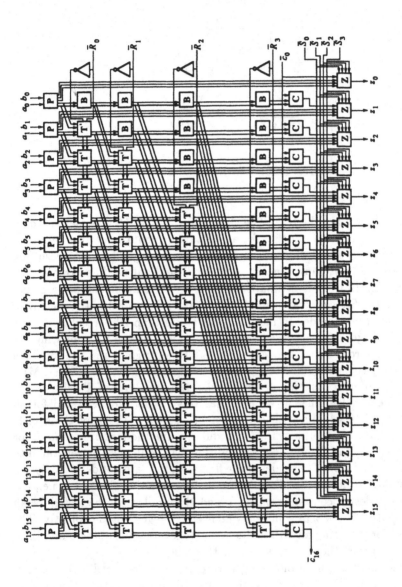

Figure 4.21: Modified tree design for ALU of Fig. 4.17.

SSL faults in the gate level circuit can be detected by tests generated for this high-level model. The remaining 48 tests needed to detect faults in the **T** modules are generated using a gate-level model containing approximately two-thirds the number of components and lines in the unmodified design, since only the first-level **T** modules need to be considered. Thus, tests detecting only 10 percent of the SSL faults need to be generated using a gate-level model, and even then the number of components and buses in the model is substantially less than those in the unmodified design. Tests for the rest of the faults can be derived from high-level models of the circuit in which the number of components and buses is less than one-tenth the number of components and buses in the gate-level model of the unmodified circuit, leading to an overall speedup of approximately 10. This clearly illustrates the usefulness of the suggested design modification scheme.

Chapter 5

CONCLUDING REMARKS

5.1 SUMMARY

We have attempted to demonstrate the usefulness of bus-oriented high-level circuit and fault models in reducing the test set size, as well as the test generation time, for large digital circuits. Our circuit and fault models represent natural generalizations of the classical gate-level models. We have presented a high-level algorithm VPODEM which generates tests for total bus faults in the high-level circuit models. This algorithm assigns vectors to buses in the high-level model, the vectors being represented and manipulated using the vector sequence notation. The circuit and fault models, as well as the test generation algorithm, reduce to classical ones if components are restricted to single gates, and bus sizes are restricted to one. This provides a truly hierarchical approach to test generation for large circuits, in that SSL faults not detected by tests generated using the high-level models can be detected using the same algorithm and gate-level models of the circuit under test. Experimental results presented for some representative general circuits show that, in almost all the cases considered, the hierarchical approach leads to test sets that are less than half the size of test sets obtained using traditional techniques alone.

Most existing test generation techniques compute tests for SSL faults in gate-level circuit models. This low-level approach implies that important structural information about the circuit like the presence of repeated subcircuits is ignored. Large portions of VLSI circuits are typically designed by interconnecting (nearly) identical cells in a structured manner. Our modeling technique utilizes information of this type to reduce the test generation effort for large circuits. Repeated subcircuits are combined into a high-level word-oriented subcircuit which is connected to the rest of the circuit via unidirectional buses, and some special components that we have introduced called fanout and merge elements.

An interesting feature of our modeling technique is the appearance of loops in the high-level model, while the gate-level model is combinational and loop-free. However, such loops are merely a result of the model construction process and do not represent true sequential behavior; hence we have called these high-level models pseudo-sequential. Test generation for regular array-like circuits is simplified considerably by using the pseudo-sequential models at the high level, since the number of components in the pseudo-sequential model of an array-like circuit is constant, in contrast to the gate-level model where the number of components is proportional to the array size.

Next, we presented a hierarchical test generation algorithm VPODEM designed for the foregoing circuit models. VPODEM performs a search of the set of possible input combinations similar to that performed by the standard PODEM algorithm. However, due to the grouping of input lines into buses, VPODEM's search procedure is inherently more efficient. The major steps in VPODEM, viz., initial objective selection, backtracing, performing implications, and decision-tree updating, are similar to those in PODEM. However, the sequence of these steps in VPODEM deviates significantly from PODEM due to the need to handle loops in pseudo-sequential circuits. Such loops may also necessitate more than one pass through the test generation process performed by VPODEM to obtain a single test pattern. These passes, which we termed pseudo-iterations, generate successive refinements of the input assignments. Backtracking in VPODEM is also different from PODEM, since it may now be performed across several pseudo-iteration steps. Our experimental data suggests that the number of pseudo-iterations can be expected to be small; it does not exceed two for any of the sample circuits considered. Experimental data also show that our hierarchical approach leads to significantly smaller test sets, and requires shorter test generation times than conventional gate-level algorithms.

Finally, we addressed design for testability, and showed that testing of some classes of irregular circuits can be simplified significantly by minor modifications to the original designs. First, ad hoc design modifications were presented for two important classes of circuits containing relatively small irregularities that make it difficult to construct efficient high-level models for the unmodified circuits. These include array-like circuits such as carry-lookahead generators and counters, and tree-like circuits such as decoders and decoding trees. The potential of ad hoc design modifications and high-level modeling in reducing test generation complexity is perhaps best illustrated by the decoder examples, where the complexity of the test generation problem is reduced from exponential to linear by adding a very small amount of control logic. A systematic design modification scheme called the level separation or LS scheme was developed for a class of circuits called (generalized) tree circuits which provide fast implementations of various important arithmetic functions. The fanout structure of a typical generalized tree circuit makes it difficult to construct a non-trivial high-level model for it. However, a simple systematic modification of the cells in the tree design that allows cells in each level to be tested separately, also allows a compact high-level model to be constructed, with which our test generation algorithm can be used very efficiently. This modification scheme was applied to the tree design of a 16-bit 4-function ALU. The resulting speedup in the test generation process is approximately 10, while the test set size is reduced by a factor of approximately 1.5.

5.2 FUTURE DIRECTIONS

We have seen that our high-level modeling and test generation techniques can substantially reduce the test generation complexity for combinational circuits, especially for structured circuits like k-regular arrays. However, further work can be done to increase the efficiency and the scope of the test generation process. Some possible directions for this are now discussed.

The VPODEM algorithm can be extended to handle high-level sequential circuits as outlined in Chapter 2. The implementation of this extension should not be difficult. It will differ from conventional gate-level test generators due to the necessity for dealing with VPODEM's pseudo-iterations as well as the true iterations arising from the sequential nature of the circuit under test.

The problem of redundancy, i.e., the presence of undetectable faults, in high-level circuit models represents another possible area of future work. Interestingly, a total bus fault in the high-level circuit model may be undetectable even though SSL faults on individual lines of the bus are all detectable. This situation is easily demonstrated using the parity checker circuit of Fig. 2.22 where the fault "bus 83 totally stuck-at-1" is not detectable, because compatibility requirements cannot be met while propagating the error signal to output bus 89. The lack of a test for this bus fault does not, however, imply that the lines of this bus are redundant in the classical sense [Fuj86]. This situation can hence be described as *pseudo-redundancy*, a type of high-level redundancy that vanishes when we look at the circuit at the gate level. An interesting open problem is to determine conditions for all pseudo-redundant faults in the high-level model of a circuit, especially an MPS model of k-regular circuit, to be detectable (or undetectable) in the corresponding low-level model. Such conditions could be usefully incorporated into VPODEM to enhance its performance.

A basic assumption of our research is that both the high and the low-level models of circuits under test, and the locations of the repeated subcircuits, are known. However, sometimes only a gate-level model of the circuit is available, and the presence of high-level structures in the circuit may not be known. In such cases, the major difficulty in using our test generation technique is constructing a suitable high-level model. This requires identifying the main repeated subcircuits in the gate-level model of the circuit. Although the general problem appears to be intractable, it may have a solution for some useful special cases. For example, this problem has been been addressed in [You86] for the case of nearly k-regular circuits, and a heuristic solution has been proposed based on identifying subcircuit isomorphism. Finding reasonably efficient heuristic techniques to solve this subcircuit isomorphism problem for more general circuits remains an interesting open problem.

The main set of logic values considered in this book can be extended beyond the conventional five values, viz., $S = \{0, 1, D, \overline{D}, X\}$. Preliminary studies show that such generalizations have the potential of increasing the efficiency of the test generation algorithm itself, especially while handling redundant circuits. It has also been shown [Hay86] that both the five-valued algebra S and the nine-valued algebra in [Mut76] can be derived from a primitive set of signal values $S' = \{0, 1\}$ using set-theoretic concepts. A much larger number of different algebras can be created if we extend the basic set of signal values from $\{0, 1\}$ to $\{0, 1, Z, U\}$, where Z and U represent the high-impedance signal, and the unknown values respectively. As discussed is Chapter 3, such extensions are necessary to handle components like tristate drivers and receivers that are frequently used in modern digital circuits. Subsets of such algebras have been used in other work [Ita86] to handle test generation for tristate buffers. However, the ad hoc nature of such efforts precludes the possibility of systematically comparing the capabilities of the various signal algebras, or of systematically searching for better ones. Our high-level test generation technique could benefit from being able to handle tristate transceivers in the same general algebraic framework as other high-level components.

Although our test generation program performs some fault simulation, we have not explicitly explored the high-level model's implications for the design of more efficient fault simulators. Its advantages for test generation should carry over to simulation as well. For example, by grouping together lines into buses, these models may make it possible to simulate repeated subcircuits simultaneously, and thus allow faster simulation than is possible with conventional methods. However, difficulties arise due to the possibility that all components connected to a bus may not be active simultaneously, so that it may be necessary to simulate signal changes on portions of a bus. A more powerful vector manipulation technique that that of VPODEM would also be helpful for this type of simulation, as well as for test generation.

Our methodology appears to be generalizable to the highest levels in the VLSI design hierarchy, all of which retain the bus-oriented structure of the register level. In such cases, more general partial bus fault models than those we have defined are likely to be needed, along with new fault models. We feel that the hierarchical circuit modeling and test generation techniques of the types presented in this book, combined with such general fault models, will lead to the powerful new test generation algorithms that will be needed for digital circuits of the future.

Appendix A

PROOFS OF THEOREMS

Complete proofs of theorems 3.2, 3.3, and 4.1 are presented in this appendix.

A.1 PROOF OF THEOREM 3.2

We begin with a restatement of the theorem, followed by an example and some useful lemmas leading to the proof of the theorem.

Theorem 3.2: Let a 1-regular array consist of modules with n input lines $x_1, \cdots, x_n$ and m output lines $f_1, \cdots, f_m$ of which the first v, viz., $f_1, \cdots, f_v$, are directly observable (vertical) output lines. Let each module implement an f_j, for $1 \leq j \leq m$, as a canonical two-level AND-OR realization of the form

$$f_j(x_1, \cdots, x_n) = \sum_{i=1}^{p} x_1^{e_{i,j,1}} \cdots \cdots x_n^{e_{i,j,n}} \tag{A.1}$$

where $e_{i,j,k} \in \{0, 1\}$, and x^1 and x^0 represent x and $\bar{x}$ respectively. Let P be the set of input patterns, designated *parallel tests*, which can be applied in parallel to every module of the 1-regular array. If tests for total bus faults only in the MPS model of the array are to detect all SSL faults, then all vectors in the sets of vectors $\{V_1'\}, \{V_2'\}, \cdots, \{V_v'\}$ constructed using the CHKVCT procedure of Fig. 3.5 must belong to the set P.

An example of an array of the above type is found in a ripple-carry adder constructed using the full-adder module shown in Fig. A.1. In this case, $m = 2$ and $v = 1$, since there are two outputs $f_1 = s$ and $f_2 = c_{i+1}$, and only f_1 output is directly observable. The canonical sum-of-products Boolean equations for s and c_{i+1} corresponding to Fig. A.1 are as follows

$$s = a^1 b^0 c_i^0 + a^0 b^1 c_i^0 + a^0 b^0 c_i^1 + a^1 b^1 c_i^1$$

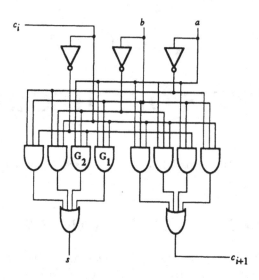

Figure A.1: Canonical realization of a full-adder module.

$$c_{i+1} = a^1 b^1 c_i^1 + a^0 b^1 c_i^1 + a^1 b^0 c_i^1 + a^1 b^1 c_i^0$$

Since a ripple-carry adder cell has only one vertical output per cell, only one set of vectors, viz., $\{V_1'\}$ is needed; its construction has already been illustrated in Fig. 3.6. We will use this example throughout this appendix to illustrate the the different steps in the proof.

Lemma A.1: A two-level AND-OR canonical realization of a function f cannot contain an undetectable s-a-0 fault on any input or output line of an AND gate.

Proof: Assume each minterm of f is implemented by only one AND gate, and let the faulty AND gate be G with k input lines. The lemma follows from the requirement that 1_k must be applied to the inputs of G to detect a s-a-0 fault on any input of G. However, since each AND gate implements a minterm of f, 1_k can be applied to the inputs of only one AND gate at a time. Hence, all AND gates other than G will have at least one 0 on their inputs and will produce 0's on their outputs. Hence, the error signal on the output of G will propagate to the output of the OR gate, causing the fault to be detected.

Lemma A.2: Let an n-input function f have the canonical form

$$f(x_1, \cdots, x_n) = \sum_{i=1}^{p} x_1^{e_{i,1}} \cdot \cdots \cdot x_n^{e_{i,n}}$$

where $e_{i,j} \in \{0,1\}$, and x^1 and x^0 represent x and $\overline{x}$, respectively, which is realized using a two-level AND-OR circuit **C**. Let G_i be the AND gate implementing the ith minterm of f. An s-a-1 fault on the jth input to G_i is undetectable if and only if

$$P_{(j)}(\overline{D}_n) \oplus ([\overline{e}_{i,1}] \odot \cdots \odot [\overline{e}_{i,n}]) \oplus ([\overline{e}_{k,1}] \odot \cdots \odot [\overline{e}_{k,n}]) = 1_n \qquad (A.2)$$

for some $k \neq i$, $1 \leq k \leq n$, where $\oplus$ denotes vector exclusive-OR.

Proof: Suppose that (A.2) is true. In order to detect the s-a-1 fault on line j of AND gate G_i, we need to apply $P_{(j)}(\overline{D}_n)$, which denotes a (column) vector of $n-1$ 1's with a 0 in the jth position, to the inputs of G_i. From the definition of minterm, applying $P_{(j)}(\overline{D}_n)$ to the inputs of gate G_i implies the application of the following vector to the inputs of **C**:

$$P_{(j)}(\overline{D}_n) \oplus ([\overline{e}_{i,1}] \odot \cdots \odot [\overline{e}_{i,n}]) \qquad (A.3)$$

For example, in Fig. A.1 G_2 generates the minterm $a.\overline{b}.\overline{c}_i$, hence $e_{2,1} = 1$, $e_{2,2} = 0$, and $e_{2,3} = 0$. It follows that applying $P_{(2)}(\overline{D}_3) = \begin{bmatrix} 1 \\ 0 \\ 1 \end{bmatrix}$ to G_2 results in applying $\begin{bmatrix} 1 \\ 0 \\ 1 \end{bmatrix} \oplus \begin{bmatrix} 0 \\ 1 \\ 1 \end{bmatrix} = \begin{bmatrix} 1 \\ 1 \\ 0 \end{bmatrix}$ to the primary inputs a, b and c_i. Furthermore, from the definition of minterm, applying the vector (A.3) to **C** implies the application of

$$P_{(j)}(\overline{D}_n) \oplus ([\overline{e}_{i,1}] \odot \cdots \odot [\overline{e}_{i,n}]) \oplus ([\overline{e}_{k,1}] \odot \cdots \odot [\overline{e}_{k,n}]) \qquad (A.4)$$

to the AND gate G_k. Referring again to Fig. A.1, applying $P_{(2)}(\overline{D}_3) = \begin{bmatrix} 1 \\ 0 \\ 1 \end{bmatrix}$ to G_2 implies applying $\begin{bmatrix} 1 \\ 0 \\ 1 \end{bmatrix} \oplus \begin{bmatrix} 0 \\ 1 \\ 1 \end{bmatrix} \oplus \begin{bmatrix} 0 \\ 0 \\ 0 \end{bmatrix} = \begin{bmatrix} 1 \\ 1 \\ 0 \end{bmatrix}$ to G_1. Hence, if (A.2) is true, then applying $P_{(j)}(\overline{D}_n)$ to G_i results in applying 1_n to G_k. Since all inputs of G_k are fault-free, a 1 is produced on the output of G_k, which blocks the error signal from G_j from propagating to the primary output line of **C**.

To prove the necessity of (A.2), we start by applying $P_{(j)}(\overline{D}_n)$ to G_i. This results in an error signal $\overline{D}$ on the output of G_i, which is an input to the only OR gate G in C. This error signal can now be stopped from propagating to the output line f only if the output of some other AND G_k gate feeding G is 1. In order to produce 1 on the output of G_k, it is necessary to apply 1_n to its inputs. However, applying $P_{(j)}(\overline{D}_n)$ to the inputs of G_i applies VS (A.4) to the inputs of G_k. Hence, the output of G_k can be 1 only if (A.2) is true. This completes the proof of the lemma.

Theorem 3.2 is proved by showing that the set $\{V_j'\}$ contains tests that must be applied to detect SSL faults on the inputs of the AND gates used in the portion of the cell that implements f_j for $1 \leq j \leq v$. It follows directly that if a vector $T \in \{V_j'\}$ does not belong to P, then a test necessary to detect some SSL fault in a module of the array cannot be applied simultaneously to all the modules of the array, i.e., cannot be a parallel test. This implies that the SSL fault in question cannot be detected by a test for some total bus fault in the array since a test for a total bus fault applies the same input pattern to all the modules. Without loss of generality, we can restrict our attention to any one vertical output, say f_1.

Case A: Assume the canonical sum-of-minterms realization of f_1 is non-redundant. To test the AND gates that implement f_1, it is necessary to apply the set of vectors $\{\overline{D}_n \cdot 1_n\}$ to their inputs, where $.$ denotes external time expansion defined in Sec. 2.1. Note that there are p AND gates in this implementation where the ith AND gate G_i implements the minterm $x_1^{e_{i,1,1}} \cdot \cdots \cdot x_n^{e_{i,1,n}}$. It can be easily verified that applying $\{\overline{D}_n \cdot 1_n\}$ to G_i implies the following vectors on the input lines of the module

$$\{(\overline{D}_n \cdot 1_n) \oplus (\overline{[e_{i,1,1}]} \odot \cdots \odot \overline{[e_{i,1,n}]})^{\cdot n+1}\} = \{(\overline{D}_n \cdot 1_n) \oplus \overline{(P_{(i)}(V_1))}^{\cdot n+1}\} \quad \text{(A.5)}$$

where $V_1 = ([e_{1,1,1}] \cdot \cdots \cdot [e_{p,1,1}]) \odot \cdots \odot ([e_{1,1,n}] \cdot \cdots \cdot [e_{p,1,n}])$. The order in which the vectors (tests) in the set described in (A.5) are applied is unimportant since the circuit under consideration is combinational. This set of vectors can also be represented as follows

$$\bigcup_{l=0}^{n} \left\{ S_{(1,..,l-1)}(P_{(i)}(V_1)) \odot \overline{S_{(l)}(P_{(i)}(V_1))} \odot S_{(l+1,..,n)}(P_{(i)}(V_1)) \right\} \quad \text{(A.6)}$$

To illustrate the equivalence of (A.5) and (A.6), consider the AND gate G_1 in the full-adder example of Fig. A.1. In this case, $e_{1,1,1} = 1, e_{1,1,2} = 1, e_{1,1,3} = 1$. Hence, (A.5) leads to the set of vectors

$$\left\{ \begin{bmatrix} 0 & 1 & 1 & 1 \\ 1 & 0 & 1 & 1 \\ 1 & 1 & 0 & 1 \end{bmatrix} \oplus \overline{\begin{bmatrix} 1 \\ 1 \\ 1 \end{bmatrix}}^{\cdot 4} \right\}$$

$$= \left\{ \begin{bmatrix} 0 & 1 & 1 & 1 \\ 1 & 0 & 1 & 1 \\ 1 & 1 & 0 & 1 \end{bmatrix} \right\}$$

$$= \left\{ \begin{bmatrix} 0 \\ 1 \\ 1 \end{bmatrix}, \begin{bmatrix} 1 \\ 0 \\ 1 \end{bmatrix}, \begin{bmatrix} 1 \\ 1 \\ 0 \end{bmatrix}, \begin{bmatrix} 1 \\ 1 \\ 1 \end{bmatrix} \right\}$$

The set of vectors derived from (A.6), on the other hand, is

$$\left\{ S_{(1..3)}(P_{(1)}(V_1)) \right\} \bigcup \left\{ \overline{S_{(1)}(P_{(1)}(V_1))} \odot S_{(2..3)}(P_{(1)}(V_1)) \right\}$$

$$\bigcup \left\{ S_{(1)}(P_{(1)}(V_1)) \odot \overline{S_{(2)}(P_{(1)}(V_1))} \odot S_{(3)}(P_{(1)}(V_1)) \right\}$$

$$\bigcup \left\{ S_{(1..2)}(P_{(1)}(V_1)) \odot \overline{S_{(3)}(P_{(1)}(V_1))} \right\}$$

$$= \left\{ \begin{bmatrix} 1 \\ 1 \\ 1 \end{bmatrix} \right\} \cup \left\{ \begin{bmatrix} 0 \\ 1 \\ 1 \end{bmatrix} \right\} \cup \left\{ \begin{bmatrix} 1 \\ 0 \\ 1 \end{bmatrix} \right\} \cup \left\{ \begin{bmatrix} 1 \\ 1 \\ 0 \end{bmatrix} \right\}$$

$$= \left\{ \begin{bmatrix} 0 \\ 1 \\ 1 \end{bmatrix}, \begin{bmatrix} 1 \\ 0 \\ 1 \end{bmatrix}, \begin{bmatrix} 1 \\ 1 \\ 0 \end{bmatrix}, \begin{bmatrix} 1 \\ 1 \\ 1 \end{bmatrix} \right\}$$

From the definitions of S_α and P_α, (A.6) can be rewritten as

$$N_{i1} = \bigcup_{l=0}^{n} \left\{ P_{(i)} \left(S_{(1,..,l-1)}(V_1) \odot \overline{S_{(l)}(V_1)} \odot S_{(l+1,..,n)}(V_1) \right) \right\}$$

N_{i1} includes all input patterns that must be applied to test the AND gate implementing the ith minterm of f_1. The set of patterns N_1 necessary to detect all SSL faults on input/output lines of all the AND gates implementing f_1 is $\bigcup_{i=1}^{p} N_{i1}$, hence

$$N_1 = \bigcup_{i=1}^{p} \bigcup_{l=0}^{n} \left\{ P_{(i)} \left(S_{(1,..,l-1)}(V_1) \odot \overline{S_{(l)}(V_1)} \odot S_{(l+1,..,n)}(V_1) \right) \right\}$$

$$= \bigcup_{l=0}^{n} \bigcup_{i=1}^{p} \left\{ P_{(i)} \left(S_{(1,..,l-1)}(V_1) \odot \overline{S_{(l)}(V_1)} \odot S_{(l+1,..,n)}(V_1) \right) \right\}$$

However, by definition of the $\bigcup$ and $P_{(i)}$ operators, $\bigcup_{i=1}^{p} \{P_{(i)}(W)\} = \{W\}$ for any VS W with p columns. Consequently,

$$N_1 = \bigcup_{l=0}^{n} \left\{ \left(S_{(1,..,l-1)}(V_1) \odot \overline{S_{(l)}(V_1)} \odot S_{(l+1,..,n)}(V_1) \right) \right\} = \{V_1'\}$$

Hence, if some $T \in \{V_1'\}$ does not belong to P, then a necessary test pattern cannot be applied in parallel to all modules of the array, which implies that a test cannot be generated for some total bus fault. Since every SSL fault in the circuit is assumed to be non-redundant, tests for total bus faults will obviously not detect all detectable SSL faults.

Case B: The theorem is now proved for the case when the module realization is redundant, i.e., some of the SSL faults in the cells are not detectable. In this case, we confine our attention to detectable SSL faults only.

Lemma A.1 states that all s-a-0 faults on the inputs of AND gates in any module are detectable. Since these faults can be detected by applying 1_n to the AND gates, the vectors in N_1 that are applied to the modules to detect the s-a-0 faults constitute the set M_1 defined below:

$$M_1 = \left\{ S_{(1..-1)}(V_1) \odot \overline{S_{(0)}(V_1)} \odot S_{(1..n)}(V_1) \right\}$$

From Lemma A.2, we know that there can be undetectable s-a-1 faults on the inputs of certain AND gates. Since we are confining our attention to detectable SSL faults only, we may delete vectors from N_1 that attempt to detect such undetectable faults. Now, s-a-1 faults on inputs of an AND gate can be detected only by the application of $\{\overline{D_n}\}$ to its inputs. Hence, the vectors in N_1 that are applied to the modules to detect s-a-1 faults constitute the set M_2 defined below:

$$M_2 = \bigcup_{l=1}^{n} \left\{ S_{(1..l-1)}(V_1) \odot \overline{S_{(l)}(V_1)} \odot S_{(l+1..n)}(V_1) \right\}$$

Lemma A.2 also implies that the input pattern applied to the module in an effort to detect an undetectable s-a-1 fault on the rth input of gate G_{i1} is

$$P_{(r)}(\overline{D_n}) \oplus ([\overline{e}_{i,1,1}] \odot \cdots \odot [\overline{e}_{i,1,n}]) = ([e_{k,1,1}] \odot \cdots \odot [e_{k,1,n}])$$

for some $k \neq i$. By definition, $([e_{k,1,1}] \odot \cdots \odot [e_{k,1,n}]) = P_{(k)}(V_1)$. This implies that the vectors in M_2 that attempt to detect undetectable s-a-1 faults are columns of V_1. To delete these vectors from N_1, it suffices to check for, and delete, the vectors that are common to M_2 and $\{V_1\}$. Let $N_1^r \subseteq N_1$ consist of the columns of N_1 that must be applied to a module to detect all detectable faults on inputs of the AND gates in a possibly redundant implementation of f_1. From the above discussion, it follows that

$$N_1^r = (M_2 - \{V_1\}) \bigcup M_1$$

where $S_1 - S_2$ denote the set theoretic difference of two sets S_1 and S_2. Therefore,

$$N_1^r = \left(\bigcup_{l=1}^{n} \left\{ S_{(1..l-1)}(V_1) \odot \overline{S_{(l)}(V_1)} \odot S_{(l+1..n)}(V_1) \right\} - \{V_1\} \right)$$
$$\bigcup \left\{ S_{(1..-1)}(V_1) \odot \overline{S_{(0)}(V_1)} \odot S_{(1..n)}(V_1) \right\}$$

However, $M_1 = \left\{ S_{(1..-1)}(V_1) \odot \overline{S_{(0)}(V_1)} \odot S_{(1..n)}(V_1) \right\} = \{V_1\}$, and $(S_1 - S_2)$ $\bigcup S_2 = S_1 \bigcup S_2$ for any two sets S_1 and S_2, implying that $N_1^r = N_1$. The theorem follows immediately.

A.2 PROOF OF THEOREM 3.3

This section begins with a restatement of the theorem. Some useful lemmas are proved next, and the theorem itself is finally proved.

Theorem 3.3 A test for a total bus fault $F_e = e_n, e \in \{0,1\}$ on a bus B in the MPS model of a k-regular circuit generated using VPODEM detects all SSL faults $f_e = e_1$ on individual lines of the bus in the corresponding gate-level model.

As stated in Chapter 3, we restrict our attention to k-regular arrays in which the signal flow between adjacent modules is unidirectional. As usual, the gate-level model of a k-regular array is denoted $\mathbf{M}^G$, and its high-level MPS model is denoted $\mathbf{M}^H$. The quasi- and pseudo-state input buses in $\mathbf{M}^H$ to which $D/\overline{D}$ error signals are assigned during pseudo-iteration step p, for $p \geq 2$, will be considered to carry a *pseudo-fault* denoted F_p^H. Moreover, we write $F_e = F_1^H$, i.e., the pseudo-fault in the first pseudo-iteration step is the total bus fault for which a test is being generated. However, in subsequent pseudo-iteration steps, only pseudo-faults on quasi- and pseudo-state input buses exist, since these are the only buses that were assigned error vectors at the end of the previous step; the original bus fault F_e is neglected for all but the first pseudo-iteration step. Corresponding to each F_p^H, we also have a set of possible gate-level pseudo-faults $\{F_{sp}^G\}$, where F_{sp}^G represents a pseudo-fault in the period s of $\mathbf{M}^G$, which consists of SSL faults on lines in the period s corresponding to the faulty buses in F_p^H. When $p = 1$, F_{s1}^G consists of the SSL fault f_e on an appropriate line in the period s. When $p \geq 2$, however, F_{sp}^G consists of faults on the horizontal input lines $X_{(s-1)k+1,j}^G$ to module $C_{(s-1)k+1}^G$ only.

Let $\mathbf{B}_p$ represent the set of output buses onto which error vectors are propagated at the end of pseudo-iteration step p. In general, $\mathbf{B}_p$ consists of vertical output buses, i.e., $\hat{Z}_{ij}^H$'s, and pseudo- and quasi-state output buses, i.e., $\hat{P}_{ij}^H$'s, and O_{ij}^H's, where the notation for the various buses is as shown in Fig. 2.16. Let V_p be the partial input assignment vector determined at the end of pseudo-iteration p. Let v_{sp} be the input assignment to the period s of modules of $\mathbf{M}^G$ implied by V_p. Let $\mathbf{L}_{sp}$ be the set of output lines of modules of $\mathbf{M}^G$ in the period s, which receive some error signal when input v_{sp} is applied to $\mathbf{M}^G$ in the presence of pseudo-fault $F_s p^G$. In general, $\mathbf{L}_{sp}$ consists of vertical output lines $\hat{Z}_{ij}^G$, and horizontal output lines $\hat{X}_{sk,j}^G$. Note that unlike $\mathbf{B}_p$, $\mathbf{L}_{sp}$ contains lines (denoted $\hat{X}_{sk,j}^G$) which are actually internal lines of $\mathbf{M}^G$. These lines are connected to the modules $C_{sk+1,j}^G$ as input lines, and hence error signals on them can be thought to define a gate-level pseudo-fault in period $s+1$ of $\mathbf{M}^G$.

The various sets defined above are illustrated using the example of test generation for the MPS model of a parity checker circuit. The low-level and the high-level (MPS) models of the circuit have already been presented in Figs. 2.13c and 2.22, respectively. For convenience, the two models are shown again in Fig. A.2. The partial input assignments generated by VPODEM during test generation for bus fault bus 55 stuck-at-0 are shown in Fig. A.3. Here, set $\mathbf{B}_1$ consists of buses 62, 63, 64, 65, 67 and 69. Note that $\mathbf{B}_1$ consists of pseudo- and quasi-state output buses only, requiring VPODEM to perform a second pseudo-iteration step. If we now assume that an SSL fault is actually present on the second line of bus 55, i.e., in the second period of the array, then set $\mathbf{L}_{21}$ consists of lines $W_{2,1}$ and $W_{2,2}$ shown in Fig. A.2, and the error signals on these lines now correspond to a pseudo-fault F_{12}^G. $\mathbf{B}_2$, on the other hand, consists of bus 89 only. This implies that $\mathbf{L}_{12}$ is now non-empty, and contains the vertical output line $Z_{2,1}$ shown in Fig. A.3, implying that an error signal has been propagated onto a directly observable line, i.e., the SSL fault will be detected by the input assignment determined by VPODEM.

From the proof outline in Section 3.2, we recall that it suffices to prove the theorem for the case when signal flow between adjacent cells in the array is unidirectional. Thus, without any loss of generality, we assume the direction of signal flow between adjacent modules to be from left to right throughout the array.

The proof of this theorem is based on the following central idea: when iden-

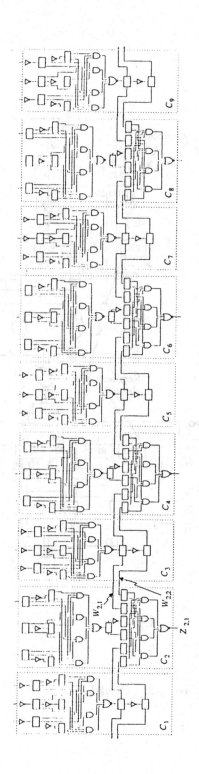

Figure A.2: Models of a parity-checker: (a) low-level model; (b) MPS model.

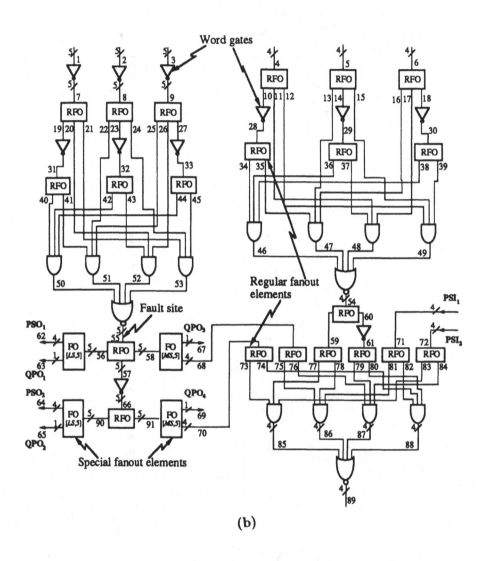

(b)

Figure A.2: (Contd.)

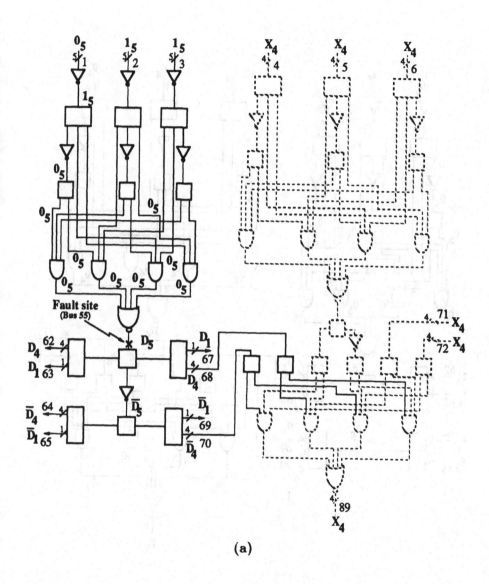

(a)

Figure A.3: Test generation using MPS model of the parity checker: (a) input assignment after the first pseudo-iteration step; (b) input assignment after the second pseudo-iteration step.

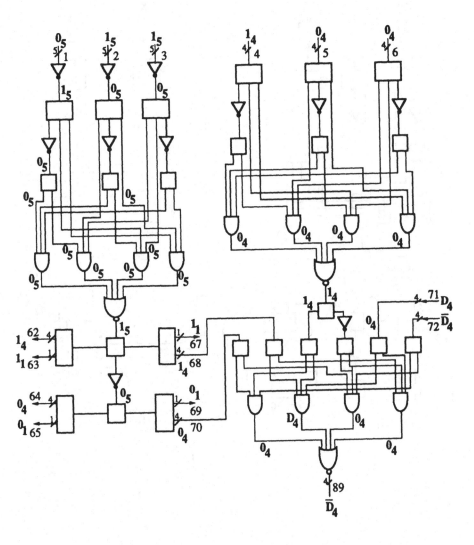

(b)

Figure A.3: (Contd.)

tical inputs can be applied to modules in periods $1 : s$ of $\mathbf{M}^G$, a pseudo-fault in period s can either lead to an error signal on a vertical (observable) output line, or to a pseudo-fault in period $s + 1$. Note that the input pattern being applied to $\mathbf{M}^G$ may not generate a error signal on any output line. However, in such cases, $\mathbf{B}_p = \phi$ at some pseudo-iteration step, which means that the test generation process will not identify such input patterns as tests.

Lemma A.3 For $1 \leq s \leq q$, there exists a one-to-one correspondence between (i) the vertical input lines $Z^G_{(s-1)k+i,j}$ in $\mathbf{M}^G$, and the lines of bus $Z^H_{i,j}$ in $\mathbf{M}^H$; (ii) the vertical output lines $\hat{Z}^G_{(s-1)k+i,j}$, and the lines of bus $\hat{Z}^H_{i,j}$; (iii) the lines $X^G_{(s-1)k+1,j}$, and the lines of the buses $P^H_{1,j}$ and $I^H_{1,j}$ taken together; (iv) the lines $\hat{X}_{sk,j}$, in $\mathbf{M}^G$, and the lines of the buses $\hat{P}^H_{n,j}$ and $O^H_{n,j}$ taken together.

Proof: These correspondences follows from the construction of $\mathbf{M}^H$ from $\mathbf{M}^G$, as formally described in Fig. 2.17. For example, the vertical bus $Z^H_{i,j}$ in $\mathbf{M}^H$ is obtained by grouping together the lines $Z^G_{i,j}, Z^G_{k+i,j}, \cdots, Z^G_{(q-1)k+i,j}$ of $\mathbf{M}^G$. The correspondence between the horizontal input lines in $\mathbf{M}^G$, and the lines in the input buses of $\mathbf{M}^H$ follows similarly, if we note that $I^H_{1,j}$ is the same as $X^G_{1,j}$, and $P^H_{1,j}$ is a renaming of the group of lines $X^G_{k+1,j}, X^G_{2k+1,j}, \cdots, X^G_{(q-1)k+1,j}$. This renaming is performed by the special merge element $\mathrm{ME}_{1j}[LS, |X^H_{1,j}|]$. The correspondence between the output lines in $\mathbf{M}^G$ and the output buses in $\mathbf{M}^H$ follows similarly.

Lemma A.4 If a pseudo-fault F^G_{sp} is present in $\mathbf{M}^G$, and the partial input assignment at the end of the p pseudo-iteration step is V_p, then $\mathbf{B}_p \neq \phi$ implies $\mathbf{L}_{sp} \neq \phi$. Moreover, $\hat{Z}^G_{(s-1)k+i,j} \in \mathbf{L}_{sp}$ if and only if $\hat{Z}^H_{ij} \in \mathbf{B}_p$. Also, $\hat{X}^G_{sk,j} \in \mathbf{L}_{sp}$ if and only if $\hat{P}^H_{nk,j} \in \mathbf{B}_p$.

Proof: Compatibility checks in TESTGEN and justification in ITERATE ensure that applying V_p to $\mathbf{M}^G$ implies the application of v_{sp} to period s in $\mathbf{M}^G$. This lemma then follows directly from Lemma A.3.

Lemma A.5 If ITERATE terminates in success after $p \geq 1$ pseudo-iteration steps, then $\mathbf{B}_1 \neq \phi, \mathbf{B}_2 \neq \phi, \cdots, \mathbf{B}_p \neq \phi$, and one of the following holds for all i and j: (i) $p = q$ and $\hat{Z}^H_{i,j} \notin \mathbf{B}_r$ for $1 \leq r \leq p$, or (ii) $p < q$, $\hat{Z}^H_{i,j} \in \mathbf{B}_p$ for some i and j, and $\hat{Z}^H_{i,j} \notin \mathbf{B}_r$ for $1 \leq r \leq p - 1$.

Proof: The only primary output buses in $\mathbf{M}^H$ are the buses $\hat{Z}^H_{i,j}$. All other output buses are designated pseudo-state output or quasi-output buses. The only conditions under which ITERATE terminates in success are when an error vector appears on a primary output bus, or when the pseudo-iteration count equals q. On the other hand, the only condition under which ITERATE enters pseudo-iteration step $r+1$ for $r \geq 1$ is $\mathbf{B}_r \neq \phi$, and $\hat{Z}^H_{i,j} \notin \mathbf{B}_r$ for all i and j. The lemma follows directly.

Lemma A.6 Let fault f_e lie in period $s+1$ of $\mathbf{M}^G$. Let V_p be the input assignment generated by ITERATE when it terminates in success after p pseudo-iteration steps. If V_p is applied to $\mathbf{M}^G$, then one of following holds: (i) $p \geq q-s$ and $\mathbf{L}_{(s+r)r} \neq \phi$ for $1 \leq r \leq q-s$, (ii) $p < q-s$ and $\mathbf{L}_{(s+r)r} \neq \phi$ for $1 \leq r \leq p$, $\hat{Z}^G_{(s+r-1)k+i,j} \notin \mathbf{L}_{(s+r)r}$ for $1 \leq r < p$ and all i,j, and $\hat{Z}^G_{(s+p-1)k+i,j} \in \mathbf{L}_{(s+p)p}$, for some i,j.

Proof: From the assumption that ITERATE terminates successfully after p pseudo-iterations, and from Lemma A.5, we conclude that $\mathbf{B}_1 : \mathbf{B}_p \neq \phi$. We need to consider the two cases that arise from Lemma A.5.

Case 1. Let $p = q$ and $\hat{Z}^H_{i,j} \notin \mathbf{B}_r$ for $1 \leq r \leq p$. Since the first gate-level pseudo-fault that occurs in $\mathbf{M}^G$ is $F^G_{(s+1)1}$, $\mathbf{B}_1 \neq \phi$ implies $\mathbf{L}_{(s+1)1} \neq \phi$. However, from Lemmas A.4 and A.5, $\hat{Z}^G_{sk+i,j} \notin \mathbf{L}_{(s+1)1}$ for all i,j. Hence, $\mathbf{L}_{(s+1)1}$ consists of horizontal lines only, and the error signals on lines in $\mathbf{L}_{(s+1)1}$ now imply pseudo-fault $F^G_{(s+2)2}$ in period $s+2$ of $\mathbf{M}^G$, since the total bus fault F_e is neglected for the pth pseudo-iteration step when $p > 1$. Following similar reasoning, $\mathbf{B}_2 \neq \phi$ and the presence of $F^G_{(s+2)2}$ implies $\mathbf{L}_{(s+2)2} \neq \phi$ (Lemma A.4). Repeating this argument r times, we conclude that $\mathbf{L}_{(s+r)r} \neq \phi$ for $1 \leq r \leq p$ implying $\mathbf{L}_{(s+r)r} \neq \phi$ for $1 \leq r \leq q$ (since $p = q$) further implying $\mathbf{L}_{(s+r)r} \neq \phi$ for $1 \leq r \leq q-s$.

Case 2. For all i and j, let $p < q$, $\hat{Z}^H_{i,j} \in \mathbf{B}_p$ for some i and j, and $\hat{Z}^H_{i,j} \notin \mathbf{B}_r$ for $1 \leq r \leq p-1$.

- **Subcase 2.1.** $p > q-s$: From Lemma A.5, $\hat{Z}^G_{(s+r-1)k+i,j} \notin \mathbf{L}_{(s+r)r}$ for $1 \leq r \leq p-1$. Since $p > q-s$, we have $\hat{Z}^G_{(s+r-1)k+i,j} \notin \mathbf{L}_{(s+r)r}$ for $1 \leq r \leq q-s$. Now by the argument of Case 1, $\mathbf{L}_{(s+r)r} \neq \phi$.

- **Subcase 2.2.** $p = q-s$: Following the argument of Subcase 2.1, $\hat{Z}^G_{(s+r-1)k+i,j} \notin \mathbf{L}_{(s+r)r}$ for $1 \leq r \leq q-s-1$. Moreover, error signals on lines in $\mathbf{L}_{(q-1)(q-s-1)}$ imply pseudo-fault $F^G_{(q-s)}$. Therefore, $\mathbf{B}_p = \mathbf{B}_{q-s} \neq \phi$ implies $\mathbf{L}_{q(q-s)} \neq \phi$ (Lemma A.4). Thus, $\mathbf{L}_{(s+r)r} \neq \phi$ for $1 \leq r \leq q-s$.

- **Subcase 2.3.** $p \leq q-s$: Again, following the argument of Subcase 2.1, $\hat{Z}^G_{(s+r-1)k+i,j} \notin \mathbf{L}_{(s+r)r}$ for $1 \leq r \leq p-1$, and error signals on lines in

$\mathbf{L}_{(s+p-1)(p-1)}$ imply pseudo-fault F_p^G. From Lemma A.4, $\hat{Z}_{i,j}^H \in \mathbf{B}_p$ now implies that $\hat{Z}_{(s+p-1)k+i,j}^G \in \mathbf{L}_{(s+p)p}$, for some i, j, and $\mathbf{L}_{(s+p)p} \neq \phi$. This completes the proof of the lemma.

The theorem follows directly from Lemma A.6. Note that if an error f_e exists in period $s + 1$ of $\mathbf{M}^G$ for $1 \leq s + 1 \leq q$, and if V_p is a test generated by ITERATE and TESTGEN for the fault F_e in $\mathbf{M}^H$, then one of the following holds when V_p is applied to $\mathbf{M}^G$: (i) if $p < q - s$, then $\mathbf{L}_{(s+p)p}$ contains a vertical output line, since $\hat{Z}_{(s+p-1)k+i,j} \in \mathbf{L}_{(s+p)p}$ for some i, j; (ii) if $p \geq q - s$ then an error signal appears at some output of the module C_{qk}^G since $\mathbf{L}_{qr} \neq \phi$ for some $r \leq p$. A vertical output line of $\mathbf{M}^G$ is an observable line, as is any output line of module qk in $\mathbf{M}^G$. Hence, f_e is detected by V_p, completing the proof of the theorem.

A.3 PROOF OF THEOREM 4.1

As in the previous sections, the theorem is restated first, and is then followed by the proof.

Theorem 4.1 Let Ω be the class of functions realizable by $m \times n$ ILA's such that the output vector of a supermodule can be described by a recurrence relation of the form

$$X_{i+1} = f_{K_i}(X_i) = f_{(A_i, B_i)}(X_i) = f(A_i, g(B_i, X_i)) \qquad (A.7)$$

where A_i and B_i are constant vectors. The functions f and g must also have the following properties:

- **P1:** f and g are associative, i.e., $f(X, f(Y, Z)) = f(f(X, Y), Z)$ and $g(X, g(Y, Z)) = g(g(X, Y), Z)$.
- **P2:** g distributes over f, i.e., $g(X, f(Y, Z)) = f(g(X, Y), g(X, Z))$.
- **P3:** $|f| = |g|$, where $|f|$ denotes the size of the output vector produced by f.

It is assumed that $|X_i + 1| = |X_i|$, i.e., $|X_i| = |f|$, and that $|X_i| = |A_i| = |B_i|$. Then there exists a generalized tree realization of each function in Ω with delay $T = O(\log_2 m \log_2 n)$ and area $A = O(mn \log_2 m \log_2 n)$. Furthermore, this generalized tree design can be modified with $O(\log_2 m \log_2 n)$ extra control signals so that it has a test set of size $S = O(\log_2 m \log_2 n)$.

Proof: We first determine the time and area requirements of computing the semigroup operation $\circ$ for the functions $f_{(A_i, B_i)}$ defined by (A.7). The supermodules are assumed to have input and output vectors of size n. (The same arguments apply if the vectors are of size m.) By direct expansion of $f_{(A_i,B_i)} \circ f_{(A_{i-1},B_{i-1})}$, and using the properties (P1) and (P2) above, we see that

$$f_{(A_i,B_i)} \circ f_{(A_{i-1},B_{i-1})}(X) = f_{(A_i,B_i)}(f_{(A_{i-1},B_{i-1})}(X))$$
$$= f(A_i, g(B_i, f(A_{i-1}, g(B_{i-1}, X)))) \text{ by definitions of } f_{(A_i,B_i)} \text{ etc.}$$
$$= f(A_i, f(g(B_i, A_{i-1}), g(B_i, g(B_{i-1}, X)))) \text{ by distributivity of } g \text{ over } f$$
$$= f(A_i, f(g(B_i, A_{i-1}), g(g(B_i, B_{i-1}), X))) \text{ by commutativity of } g$$
$$= f(f(A_i, g(B_i, A_{i-1})), g(g(B_i, B_{i-1}), X)) \text{ by commutativity of } f$$
$$= f(A_j, g(B_j, X)) = f_{(A_j,B_j)}(X)$$

where $A_j = f(A_i, g(B_i, A_{i-1}))$, and $B_j = g(B_i, B_{i-1})$. It follows from the definition of the supermodules that there exists a one-dimensional ILA of size n that implements $f(A, g(B, C))$. Moreover, the construction of the supermodules only requires $|f| = |X_i|$. By property (P3), $|f| = |g|$, therefore we can replace f by the function $f_I(X, Y) = Y$ and still implement f_I by a one-dimensional ILA of size n. This implies that $g(B, C)$ can also be implemented by the same type of ILA. Hence, $f_{(A_j,B_j)}$ in equation (A.7) can be computed from $f_{(A_i,B_i)}$ and $f_{(A_{i-1},B_{i-1})}$ using a one-dimensional ILA of size n. However, from [Abr80] we know that such ILA's can be replaced by tree circuits with delay $O(\log_2 n)$ and area $O(n \log_2 n)$. The semigroup operation for the vector functions $f_{(A_i,B_i)}$ can therefore be computed by a supertree (**ST**) module in time $O(\log_2 n)$ and area $O(n \log_2 n)$.

Now, we can replace the one-dimensional ILA of supermodules by a tree circuit with $O(m \log_2 m)$ **ST** modules and $O(\log_2 m)$ **ST** delays, where an **ST** module computes the semigroup operation of the vector functions $f_{(A_i,B_i)}$. Combining these tree parameters with those of the **ST** modules noted above, the total area A of the final circuit is seen to be $O(mn \log_2 m \log_2 n)$, and its total delay T is $O(\log_2 m \log_2 n)$. The size of the test set and the number of control signals follow from the fact that the final circuit is a generalized tree-type circuit with $O(\log_2 m)$ levels of **ST** modules, and each **ST** module contains a similar tree circuit with $O(\log_2 n)$ levels of elementary modules.

BIBLIOGRAPHY

[Aba86] M. S. Abadir and H. K. Reghbati, "Functional test generation for digital circuits using binary decision diagrams," *IEEE Trans. on Computers*, vol. C-35, July 1986, pp. 375-379.

[Abo83] M. E. Aboulhamid and E. Cerny, "A class of test generators for built-in testing," *IEEE Trans. on Computers*, vol. C-32, Oct. 1983, pp. 957-959.

[Abr80] J. A. Abraham and D. D. Gajski, "Easily testable, high-speed realization of register-transfer-level operations," *Proc. 10th Fault-Tolerant Computing Symp.*, Oct. 1980, pp. 339-344.

[Abr83] J. A. Abraham, "Design for testability," *Proc. Custom Integrated Circuits Conf.*, 1983, pp. 278-283.

[Abr84] M. Abramovici, P. R. Menon and D.T. Miller, "Critical path tracing," *IEEE Design and Test*, Feb. 1984, pp. 83-92.

[Adv77] Advanced Micro Devices Inc., *Schottky and Low-power Schottky Data Book*, Sunnyvale, Calif., 1977.

[Agr75] P. Agrawal and V. D. Agrawal, "Probabilistic analysis of random test generation method for irredundant combinational logic networks," *IEEE Trans. on Computers*, vol. C-24, July 1975, pp. 691-695.

[Agr78] V. D. Agrawal, "When to use random testing," *IEEE Trans. on Computers*, vol. C-27, Nov. 1978, pp. 1054-1055.

[Ake85] S. B. Akers, "On the use of linear sums in exhaustive testing," *Proc. 15th Fault-Tolerant Computing Symposium*, June 1985, pp. 148-153.

[And80] H. Ando, "Testing VLSI with random access scan," *Digest of Papers, Spring Compcon '80*, Feb. 1980, pp. 50-52.

[Avi75] A. Avizienis, "Fault-tolerant systems," *IEEE Trans. on Computers*, vol. C-25, Dec. 1975, pp. 1304-1312.

[Bar80]   Z. Barzilai, J. Savir, G. Markowsky and M. G. Smith, "Syndrome-testable design of combinational circuits," *IEEE Trans. on Computers*, vol. C-29, June 1980, pp. 442-451.

[Bar81]   Z. Barzilai, J. Savir, G. Markowsky and M. G. Smith, "The weighted syndrome sums approach to VLSI testing," *IEEE Trans. on Computers*, vol. C-30, Dec. 1981, pp. 996-1000.

[Bar83]   Z. Barzilai, D. Coppersmith, A. L. Rosenberg, "Exhaustive generation of bit patterns with application to VLSI self-testing," *IEEE Trans. on Computers*, vol. C-32, Feb. 1983, pp. 190-194.

[Ben84]   R. G. Bennetts, *Design of Testable Logic Circuits*, Addison-Wesley, London, 1984.

[Bha85]   D. Bhattacharya and J. P. Hayes, "High-level test generation using bus faults," *Proc. 15th Fault-Tolerant Computing Symp.*, June 1985, pp. 65-70.

[Bha86]   D. Bhattacharya and J. P. Hayes, "Fast and easily testable implementation of arithmetic functions," *Proc. 16th Fault-Tolerant Computing Symp.*, July 1986, pp. 324-329.

[Bha88]   D. Bhattacharya, *Hierarchical Modeling and Test Generation for Digital Circuits*, Ph. D. Dissertation, Program in Computer, Information, and Control Engineering, The University of Michigan, May 1988.

[Bre76]   M. A. Breuer and A. D. Friedman, *Diagnosis and Reliable Design of Digital Systems*, Computer Science Press, Woodland Hills, Calif., 1976.

[Bre80]   M. A. Breuer and A. D. Friedman, "Functional level primitives in test generation," *IEEE Trans. on Computers*, vol. C-29, March 1980, pp. 223-235.

[Bre80]   R. P. Brent and H. T. Kung, "The chip complexity of binary arithmetic," *Proc. 12th ACM Symp. on Theory of Computing*, April 1980, pp. 190-200.

[Cas76]   G. R. Case, "Analysis of actual fault mechanisms in CMOS logic gates," *Proc. 13th Design Automation Conf.*, June 1976, pp. 265-270.

[Cha88]   T. Chakrabarty and S. Ghosh, "On behavior fault modeling for combinational digital designs," *Porc. International Test Conf.*, Sept. 1988, pp. 593-600.

[Che84] C. L. Chen, "Error-correcting codes for semiconductor memories," *Proc. International Symp. on Computer Architecture*, June 1984, pp. 245-247.

[Che86] W.-T. Cheng and J. H. Patel, "Testing in two-dimensional iterative logic arrays," *Proc. 16th Fault-Tolerant Computing Symp.*, July 1986, pp. 76-83.

[Dae81] W. Daehn and J. Mucha, "A hardware approach to self-testing of large programmable logic arrays," *IEEE Trans. on Computers*, vol. C-30, Nov. 1981, pp. 829-833.

[Eic78] E. B. Eichelberger and T. W. Williams, "A logic design structure for LSI testability," *Journal of Design Automation and Fault Tolerant Computing*, May 1978, pp. 165-178.

[Eld59] R. D. Eldred, "Test routines based on symbolic logic statements," *Journal of ACM*, vol. 6, March 1959, pp. 33-36.

[Fer88] F. J. Ferguson and J. P. Shen, "Extraction and simulation of realistic CMOS faults using inductive fault analysis," *Proc. International Test Conf.*, Sep. 1988, pp. 475-484.

[Feu88] R. J. Feugate, Jr., and S. M. McIntyre, *Introduction to VLSI Testing*, Prentice-Hall, Englewood Cliffs, N. J., 1988.

[Fri73] A. D. Friedman, "Easily testable iterative systems," *IEEE Trans. on Computers*, vol. C-22, Dec. 1973, pp. 1061-1064.

[Fuj83] H. Fujiwara and T. Shimono, "On the acceleration of test generation algorithms," *Proc. 13th Fault-Tolerant Computing Symp.*, June 1983, pp. 98-105.

[Fuj86] H. Fujiwara, *Logic Testing and Design for Testability*, MIT Press, Cambridge, Mass., 1986.

[Fun75] S. Funatsu, N. Wakatsuki and T. Arima, "Test generation systems in Japan," *Proc. 12th Design Automation Symp.*, June 1975, pp. 114-122.

[Gaj77] D. D. Gajski, "Semigroup of recurrences" in D. J. Kuck et al. (eds.) *High-speed Computer and Algorithm Organization*, 1977, pp. 179-183.

[Gho88] S. Ghosh, "Behavior-level fault simulation," *IEEE Design and Test*, vol. 5, June 1988, pp. 31-42.

[Goe81] P. Goel, "An implicit enumeration algorithm to generate tests for combinational logic circuits," *IEEE Trans. on Computers*, vol. C-30, March 1981, pp. 215-222.

[Has83] S. Z. Hassan and E. J. McCluskey, "Testing PLA's using multiple parallel signatures," *Proc. 13th Fault-Tolerant Computing Symp.*, June 1983, pp. 422-425.

[Hay74] J. P. Hayes and A. D. Friedman, "Test point placement to simplify fault detection," *IEEE Trans. on Computers*, vol. C-23, July 1974, pp. 727-735.

[Hay80] J. P. Hayes, "A calculus for testing complex digital systems," *Proc. 10th Fault-Tolerant Computing Symp.*, Oct. 1980, pp. 115-120.

[Hay86] J. P. Hayes, "Uncertainty, energy, and multiple-valued logics," *IEEE Trans. on Computers*, vol. C-35, Feb. 1986, pp. 107-114.

[Hay87] J. P. Hayes, "An introduction to switch-level modeling," *IEEE Design and Test*, vol. 4, Aug. 1987, pp. 18-25.

[Hay88] J. P. Hayes, *Computer Architecture and Organization*, 2nd ed., McGraw-Hill, New York, 1988.

[Hen61] F. C. Hennie, *Iterative Arrays of Logical Circuits*, MIT Press, Cambridge, Mass., 1961.

[Hir81] S. S. Hirschhorn, M. Hommel, and C. Bures, "Functional level simulation in FANSIM3," *Proc. 18th Design Automation Conf.*, June 1981, pp. 248-255.

[Iba75] O. H. Ibarra, and S. K. Sahni, "Polynomially complete fault detection problems," *IEEE Trans. on Computers*, vol. C-24, March 1975, pp. 242-249.

[Ita86] N. Itazaki and K. Kinoshita, "Algorithmic generation of test patterns for circuits with tri-state modules," *Proc. 16th Fault-Tolerant Computing Symp.*, July 1986, pp. 64-69.

[Kau67] W. H. Kautz, "Testing for faults in combinational cellular logic arrays," *Proc. 8th Symp. on Switching and Automata Theory*, Oct. 1967, pp. 161-174.

[Koe79] B. Koenemann, J. Mucha, and G. Zwiehoff, "Built-in logic block observation techniques," *Proc. International Test Conf.*, 1979, pp. 37-41.

[Koh78] Z. Kohavi, *Switching and Finite Automata Theory*, 2nd ed., McGraw-Hill, New York, 1978.

[Kub68] H. Kubo, "A procedure for generating test sequences to detect sequential circuit failures," *NEC Journal of Research and Development*, 1967, pp. 69-78.

[Lal85] P. K. Lala, *Fault Tolerant and Fault Testable Hardware Design*, Prentice-Hall, Englewood Cliffs, N. J., 1985.

[Lee76] S. C. Lee, "Vector boolean algebra and calculus," *IEEE Trans. on Computers*, vol. C-25, Sept. 1976, pp. 865-874.

[Lev82] Y. H. Levendel and P. R. Menon, "Test generation algorithms for computer hardware description languages," *IEEE Trans. on Computers*, vol. C-31, July 1982, pp. 577-587.

[Lev83] Y. H. Levendel and P. R. Menon, "The *-algorithm: Critical traces for functions and CHDL constructs," *Proc. 13th Fault-Tolerant Computing Symp.*, June 1983, pp. 90-97.

[Lit80] M. El-Lithy and R. Husson, "Bit-sliced microprocessors testing - a case study," *Digest of Papers, IEEE Test Conf.*, Oct. 1980, pp. 126-128.

[McC80] E. J. McCluskey and S. Bozorgui-Nesbat, "Design for autonomous test," *Digest of Papers, IEEE Test Conf.*, Oct. 1980 pp. 15-21.

[McC82] E. J. McCluskey, "Verification Testing," *Digest of Papers, IEEE Test Conf.*, Nov. 1982, pp. 183-190.

[Mea80] C. Mead and L. Conway, *Introduction to VLSI Systems*, Addison-Wesley, Reading, Mass., 1980.

[Mor85] R. D. Mori and R. Cardin, "A recursive algorithm for binary multiplication and its implementation," *ACM Trans. on Computing Systems*, Nov. 1985, pp. 294-314.

[Mut76] P. Muth, "A nine-valued circuit model for test generation," *IEEE Trans. on Computers*, vol. C-25, June 1976, pp. 630-636.

[Put71] G. R. Putzolu, and J. P. Roth, "A heuristic algorithm for testing of asynchronous circuits," *IEEE Trans. on Computers*, vol. C-20, Feb. 1971, pp. 181-187.

[Rog87] W. A. Rogers, J. F. Guzolek, and J. Abraham, "Concurrent hierarchical fault simulation," *IEEE Trans. on CAD*, vol. 6, Sept. 1987, pp. 848-862.

[Rot66] J. P. Roth, "Diagnosis of automata failures: a calculus and a method," *IBM Journal of Res. and Develop.*, Oct. 1966, pp. 278-281.

[Rot80] J. P. Roth, *Computer Logic, Testing and Verification*, Computer Science Press, Potomac, Md., 1980.

[Sav78] J. Savir and P. H. Bardell, "Efficiency of random compact testing," *IEEE Trans. on Computers*, vol. C-27, June 1978, pp. 516-525.

[Sav83] J. Savir, G. Ditlow and P. H. Bardell, "Random pattern testability," *Proc. 13th Fault-Tolerant Computing Symp.*, June 1983, pp. 80-89.

[She84] J. P. Shen and F. J. Ferguson, "The design of easily testable VLSI multiplier," *IEEE Trans. on Computers*, vol. C-33, June 1984, pp. 554-557.

[Sie82] D. P. Siewiorek and R. S. Swarz, *The Theory and Practice of Reliable System Design*, Digital Press, Bedford, Mass., 1982.

[Som85] F. Somenzi, S. Gai, M. Mezzalama and P. Prinetto, "Testing strategy and technique for macro-based circuits," *IEEE Trans. on Computers*, vol. C-34, Jan. 1985, pp. 85-89.

[Sri81a] T. Sridhar and J. P. Hayes, "Design of easily testable bit-sliced systems," *IEEE Trans. on Computers*, vol. C-30, Nov. 1981, pp. 842-854.

[Sri81b] T. Sridhar and J. P. Hayes, "A functional approach to testing bit-sliced microprocessors," *IEEE Trans. on Computers*, vol. C-30, August 1981, pp. 563-571.

[Ste77] J. H. Stewart, "Future testing of large LSI circuit cards," *Digest of Papers, Semiconductor Test Symp.*, Oct. 1977, pp. 6-17.

[Tan76] A. S. Tanenbaum, *Structured Computer Organization*, Prentice-Hall, Englewood Cliffs, N. J., 1976.

[Tan83] D. T. Tang and L. S. Woo, "Exhaustive test generation with constant weight vectors," *IEEE Trans. on Computers*, vol. C-32, Dec. 1983, pp. 1145-1150.

[Tex85] Texas Instruments Inc., *The TTL Data Book*, vol. 2, Dallas, Tex., 1985.

[Tha78] S. M. Thatte and J. A. Abraham, "A methodology for functional level testing of microprocessors," *Proc. 8th Fault-Tolerant Computing Symp.*, June 1978, pp. 90-95.

[Tha80] S. M. Thatte and J. A Abraham, "Test generation for microprocessors," *IEEE Trans. on Computers*, vol. C-29, June 1980, pp. 429-441.

[Tho71] J. J. Thomas, "Automated diagnostic test generation programs for digital networks," *Computer Design*, August 1971, pp. 63-67.

[Tur79] C. Turcat and A. Verdillon, "Symmetry, automorphism and test," *IEEE Trans. on Computers*, vol. C-27, April 1979, pp. 319-325.

[Ung77] S. Unger, "Tree realizations of iterative circuits," *IEEE Trans. on Computers*, vol. C-26, April 1977, pp. 365-383.

[Ver86] A. Vergis and K. Steiglitz, "Testability conditions for bilateral arrays of combinational cells," *IEEE Trans. on Computers*, vol. C-35, Jan. 1986, pp. 13-22.

[Wai85] J. A. Waicukauski, E. B. Eichelberger, D. O. Forlenza, E. Lindbloom, and T. McCarthy, "A statistical calculation of fault detection probabilities by fast fault simulation," *Proc. International Test Conf.*, Nov. 1985, pp. 779-784.

[Wal64] C. S. Wallace, "A suggestion for a fast multiplier," *IEEE Trans. on Electronic Computers*, vol. C-13, Feb. 1964, pp. 14-17.

[You86] Y. You, *Self-testing VLSI circuits*, Ph. D. Dissertation, Dept. of Electrical Engineering and Computer Science, The University of Michigan, 1986.

[You88] Y. You and J. P. Hayes, "Implementation of VLSI self-testing by regularization," *IEEE Trans. on Computer-Aided Design*, vol. 7, Dec. 1988, pp. 1261-1271.

Index